CHARACTERIZING PEDAGOGICAL FLOW

CHARACTERIZING PEDAGOGICAL FLOW

An Investigation of Mathematics and Science Teaching in Six Countries

William H. Schmidt
Doris Jorde
Leland S. Cogan
Emilie Barrier
Ignacio Gonzalo
Urs Moser
Katsuhiko Shimizu
Toshio Sawada
Gilbert A. Valverde
Curtis McKnight
Richard S. Prawat
David E. Wiley
Senta A. Raizen
Edward D. Britton
Richard G. Wolfe

from

The Survey of
Mathematics and Science Opportunities (SMSO)

KLUWER ACADEMIC PUBLISHERS
DORDRECHT / BOSTON / LONDON

A C.I.P. Catalogue record for this book is available from the Library of Congress.

ISBN 0-7923-4272-0 (Hb)
ISBN 0-7923-4273-9 (Pb)

Published by Kluwer Academic Publishers,
P.O. Box 17, 3300 AA Dordrecht, The Netherlands.

Kluwer Academic Publishers incorporates
the publishing programmes of
D. Reidel, Martinus Nijhoff, Dr W. Junk and MTP Press.

Sold and distributed in the U.S.A. and Canada
by Kluwer Academic Publishers,
101 Philip Drive, Norwell, MA 02061, U.S.A.

In all other countries, sold and distributed
by Kluwer Academic Publishers Group,
P.O. Box 322, 3300 AH Dordrecht, The Netherlands.

Printed on acid-free paper

Printed in the Netherlands

Dedication

Leigh Burstein was a University of California-Los Angeles faculty member for 19 years. He was a principal contributor to the National Center for Research on Evaluation, Standards, and Student Testing (CRESST). He was also a participant in the IEA Second International Mathematics Study (SIMS) and editor of The IEA Study of Mathematics III: Student Growth and Classroom Processes published by Pergamon Press in 1992. He was one of the original members of the SMSO team and played a major role as a member of the US National Steering Committee for TIMSS. He deeply cared about international studies. Perhaps his most important role in TIMSS was as our personal advisor and conscience. He never gave up on the issues he felt were important, yet his support was unrelenting.

The SMSO authors and research team wish to dedicate this volume to the memory of Leigh Burstein.

A Personal Note:

I was extremely saddened over the death of my colleague, Leigh Burstein. Leigh was a member of the U.S. Steering Committee for TIMSS and a member of the SMSO research team but most importantly, he was my friend. His contributions to this study were innumerable and his loss is simply not possible to put into words. He was in Annapolis, Maryland attending a meeting of the US Steering Committee for TIMSS at the time of his death. My life was enriched and changed by his presence and I miss him terribly. This book is one small tribute to his memory.

Bill Schmidt

Contents

Preface:
The SMSO Story

Over a period of more than four years, from 1991 to present, a group of researchers from six countries have been working together in a collaborative effort to understand the key elements of teaching and learning as it occurs in mathematics and science classrooms. The goal was to bring this understanding to the development of research instruments that could be used in a subsequent large-scale investigation of the same. The original mission of the Survey of Mathematics and Science Opportunities (SMSO) was to develop a theoretical model of the educational experiences provided students and to develop a comprehensive battery of survey instruments addressing student, teacher, school, and curriculum factors. These would be used to inform the explanation and understanding of cross–national differences in student achievement in the anticipated Third International Mathematics and Science Study (TIMSS) to be conducted by the International Association for the Evaluation of Educational Achievement (IEA). This developmental research was conducted in a subset of six nations participating in TIMSS.

This book was not one of the originally envisioned products of SMSO research activity. However, as we worked together exchanging experiences with and insights into our respective educational systems we wanted to document not only what we had learned but how we had worked together. A small, international, multidisciplinary group working together to develop shared meanings and understandings about schools, schooling, and classrooms is a valuable methodological approach that can be used in developing large scale international surveys.

Early in the developmental stage, it became clear during the international research group's discussions that the assumptions brought to investigations of classrooms and what occurs within them were not always the same. The questions that seemed relevant for researchers working in one country to ask did not necessarily seem relevant or even make sense to those who operated within another country. In order to develop questionnaires that would meaningfully assess relevant factors across a wide variety of countries it was necessary to develop a common language and understanding of classrooms in the various countries.

Thus, SMSO launched a classroom observation project designed to address the question, "What does a typical mathematics (or science) lesson look like?" in each of the participating countries. Across a two year period from 1991 to 1993, researchers conducted over 120 classroom observations in France, Japan, Norway, Spain, Switzerland, and the United States. These observations were concentrated at the two student age groups that were the original focus of TIMSS, nine year-olds and thirteen year-olds. Both mathematics and science classrooms were visited in each country and these observations were written up according to an agreed-upon protocol. These classroom observations became the source of a great deal of discussion, analysis, and insight at regular meetings of the international research team.

This volume focuses upon what occurs in mathematics and science classrooms. In many ways, the classroom is like the main setting for a story that unfolds through the interactions of two types of characters: teachers and students. While teachers play an undeniably vital and significant role in these interactions, what occurs in the classroom has enormous and lifelong consequences for students. As potential educational goals are realized in actual learning experiences, students traverse invisible corridors of opportunity leading to a variety of roles for participating in and contributing to a nation's culture and society.

Chapter One elaborates on the collaborative process we employed in developing a common perspective for investigating the story of curriculum and pedagogy that occurs in classrooms. It explains in some detail how we have worked and the key role the international meetings have played in all of SMSO's endeavors. Chapter Two explores the curricular line of the story by highlighting a few of the differences in mathematics and science curricular guides and textbooks among the six countries. These differences were documented through the use of the TIMSS Curriculum Frameworks and curriculum analysis procedures.

Chapter Three presents some of the conclusions we reached about the story of classroom instruction based on the observations. Chapter Four provides examples that illustrate how some of the ideas and insights generated from the observational work can be incorporated into large scale survey instruments. Chapter Five presents some conclusions and recommendations for interpreting the results of international comparisons of student achievement. In Part II of this volume, generalized case studies set the stage for a more complete under-standing of the story as it is played out in each of the six countries. These case studies were prepared under the direction of the project coordinator in each country. They present invaluable contextual insight into each country that proved vital.

The significance of the tangible results notwithstanding, perhaps the most valuable and rewarding fruits of the enterprise occurred on a more personal level for those who participated. In the course of our discussions about math-ematics and science classrooms and what goes on in them we all experienced occasions in which our sometimes unrecognized assumptions were chal-lenged. As uncomfortable and, at times, frustrating as these occasions could be, they ultimately proved to be among the most satisfying. Indeed, there almost seemed to be a direct relationship between the frustration level pro-duced by these occasions and the value of the insight that came from resolv-ing differences and arriving at consensus. In some respects, this is one project many almost regret to have completed. Each SMSO participant came away from the project with a profound admiration and respect for the others. The stimulation and warmth of the interdisciplinary and multinational collegial effort toward a common goal is a rarity. The tangible and personal rewards will continue to have an impact for some time to come.

Acknowledgments

We gratefully acknowledge the invaluable contributions many have made to the research effort and discussion presented in this volume. Daniel Robin from France, Svein Lie from Norway, Guillermo Gil, José Antonio López Varona, and Alejandro Tiana from Spain, and Erich Ramseier from Switzerland facilitated the research effort within their respective countries. In addition to the authors, others also conducted teacher interviews, classroom observations and participated in our developing dialogue during meetings held in their country. Those who contributed in this manner include Michèle Bocquet, Jean Boudarel, Josette Le Coq, and Liliane Ruer from France, Masao Miyake and Eizo Nagasaki from Japan, Astrid Eggen Knutsen, Tone Nergård, and Torunn Nilssen from Norway, Icíar Eraña, Diana García Corona, Reyes Hernández, and Blanca Valtierra from Spain, Erich Ramseier from Switzerland, and Carol Crumbaugh, Pam Jakwerth, Mary Kino, and Margaret Savage from the United States.

At the beginning of the project, the US National Center for Education Statistics (NCES) sponsored a series of focal groups that identified issues for the instruments SMSO developed. These groups made a substantial contribution to the work present in this volume and deserve recognition. The focal group concentrating on system level characteristics was chaired by David Wiley (United States) and included Manfred Lehrke (Germany), David Stevenson (United States), Ian Westbury (United States), and Timothy Wyatt (Australia). The School Questionnaire focal group was chaired by Andrew Porter (United States) and consisted of Ray Adams (Australia), David Baker (United States), Ingrid Munck (Sweden), and Timothy Wyatt (Australia). The Teacher Questionnaire focal group was co-chaired by Leigh Burstein and Richard Prawat (United States) and included Ginette DeLandshere (Belgium), Jong-Ha Han (Korea), Mary Kennedy (United States), Fredrick K. S. Leung (Hong Kong), Eizo Nagasaki (Japan), and Teresa Tatto (Mexico). The Student Questionnaire focal group was chaired by Judith Torney-Purta (United States) and included Siew Chan (Singapore), Lois Peak (United States), Jack Schwille (United States), and Peter Vari (Hungary).

Many others in each of the participating countries were involved in key supporting roles in typing observations, translating reports, and performing the many necessary tasks required. We are especially grateful to Jacqueline Babcock whose scrupulous notes chronicled the dialogue and progress made during international meetings of the SMSO research team. Her meticulous management of all logistics for these meetings enabled our fruitful and productive dialogues – truly she was the glue that held us together. Leonard Bianchi, Christine DeMars, and Richard Houang provided invaluable assistance in organizing and analyzing data. We want to thank the National Science Foundation (NSF) who supported the SMSO research through grant SED 9054619 with funds provided by the National Center for Education Statistics. In particular, we'd like to acknowledge Larry Suter at NSF who provided supportive monitoring throughout the course of the project and Jeanne Griffith, Eugene Owen, and Lois Peak at NCES for their support and advice. As always, the opinions, findings and conclusions expressed in this publication are those of the authors and do not necessarily reflect the views of the National Science Foundation or of the National Center for Education Statistics.

Chapter 1

Investigating the Story of Curriculum and Pedagogy: Conceptualizing and Comparing Educational Practices

A teacher in an eighth grade mathematics class in Spain asks for the students' attention and announces that there are exercises to correct. Requesting numbers from some students' homework, the teacher recreates an exercise on the blackboard and reviews the concepts of positive and negative slope. As students correct their homework the teacher also asks a few questions to gauge students' comprehension.

Completing the correction of homework, the teacher tells students to turn to the appropriate textbook page and read along with him. Again, he draws on the blackboard to illustrate directly proportional linear functions. The teacher points out that students have seen this type of equation in their science class. He illustrates by drawing on the blackboard a physics example regarding the velocity of a vehicle.

Similarly, the concept of inverse proportionality is introduced, explained and illustrated with examples. Students are asked several times whether an illustration is an example of direct or inverse proportionality. After considerable explanation and discussion, the teacher writes an equation on the blackboard and asks students to copy and make a representation of it in their notebooks. He moves around the room observing students' work occasionally addressing the class to correct common errors. As the lesson concludes, the teacher assigns homework for the next lesson and reminds students of an upcoming quiz.

A teacher in Switzerland begins his eighth grade mathematics class by asking students to gather in a circle in the middle of the room. He appoints one student secretary and explains that they are going to construct two-sided stairs out of little blocks and examine the relationship between the stairs' height and the number of blocks used. Different students are asked to create stairs and the secretary records the height and number of blocks involved each time. After several sets of stairs have been constructed, the teacher asks students what the general rule is. The general relationship is summarized as $n \rightarrow n^2$ on the blackboard. One more set of stairs is built to confirm the accuracy of this relationship. Students then build stairs to discover the relationship between the stairs' height and the number of surfaces on which one could step and generalize the relationship.

Next the teacher has students physically form various polygons and count the number of diagonals. In groups of six, with students as the corners of a hexagon, string is used to form all the diagonals. After several different polygons have been constructed this way, the teacher leads a class discussion to discover the relationship between the number of sides of a polygon and the number of diagonals and to summarize this in a general formula. Several ideas are suggested and tested before one is finally adopted.

The teacher concludes the class by asking students to return to their seats and open their exercise books. He assigns a number of problems that will be due in two days. The next day they will have the option to work on any task they choose so this assignment may be completed at home or at school.

These two narratives share many commonalities: both involve mathematics, both involve students and teachers, and both tell a story of students and teachers interacting around specific curricular topics. The stories are commonplace, familiar to students and teachers alike.

In classrooms around the world, similar stories involving curriculum and pedagogy are regularly created by the actions of teachers and students. These commonplace stories have a common goal. It is this goal that reflects the importance of these stories. The goal is to produce life-long changes in students. Curriculum and pedagogy are woven together in classrooms to guide and aid students' learning. New understandings and new competencies develop as students learn. These create new opportunities for students, equipping them to play new and important roles in their society.

What occurs in the classroom is ordinary and familiar. The result is dramatic. The cumulative effect of everyday classroom experiences is similar to that of falling snow. No single snowflake or lesson makes an obvious difference; the cumulative effect is undeniable. Further, as those who dwell in the colder regions can attest, "snow" is not simply "snow". Different types of snow yield qualitatively different effects. Tiny, dry, sparkling snow flakes can effortlessly be swept away. Large, moist, heavy flakes cling to branches and require great effort to move. Lessons similarly differ qualitatively in how curriculum and pedagogy are woven together in the classroom.

One explanation for this phenomena is that teaching is fundamentally embedded in culture — both in its conception and execution. The fact that many languages have several different verbs for "teaching" underscores this notion. For example, in German, one verb refers to "teaching" at the primary or secondary level ('unterrichten') and another refers to teaching at the university level ('lehren').

What is taught is also embedded in culture. Few would likely doubt that teaching language, history, or civics is highly influenced by culture. Countries are expected to have unique perspectives on these subjects. However, most would probably consider school mathematics and science relatively unaffected by such cultural influences. Few would expect, for example, a "French" physics or a "Norwegian" mathematics. That common expectation is false. In each country, culture influences mathematics and science teaching in a way that is comparable to the teaching of language and history.

Cross-national research provides an opportunity to better understand the cumulative effects of the different ways curriculum and pedagogy are woven together in classrooms. By investigating what goes on in other classrooms, our own common, culturally influenced, and often unexamined, practices can be brought more sharply into focus. The familiar and assumed can be challenged, and a new perspective provided on common, typical practices.

This book describes the evolution of a research method and an approach to the development of instruments for use in a large scale investigation of mathematics and science classroom practices. It reports a relatively small-scale developmental study in six countries. During this study, conceptual models were developed informing the creation of the survey instruments for a larger scale study, and an approach emerged for conducting cross-national educational investigations of curriculum and pedagogy.

THE SMSO PROJECT

The Survey of Mathematics and Science Opportunities (SMSO) was the project responsible for developing instruments for the Third International Mathematics and Science Study (TIMSS). Its origin can be traced back more than five years to the late 1980's. The International Association for the Evaluation of Educational Achievement (IEA) decided then to sponsor a third large cross-national study comparing the achievement of students in mathematics. This multinational investigation was to be conducted with two student populations, nine- and thirteen-year-olds. In 1990 the IEA decided to make this new study a coordinated investigation of mathematics and science education thus changing what would have been TIMS into TIMSS, the Third International Mathematics and Science Study. Soon IEA decided to add investigating a third student population to the original two. This third population was defined as students in their final year of upper secondary education.

The final TIMSS design thus focuses on three student populations. Population One is students in the two adjacent grades in each country containing most nine-year-olds. Population Two is those two adjacent grades containing most thirteen-year-olds. Population Three is those in their last year of upper secondary school. For Populations One and Two, the main focus is upon the upper of the two adjacent grades.

In early informal discussions about the upcoming international study, representatives from a number of countries expected to participate, expressed interest in gathering data related to process as well as outcome. This request was not new. In previous studies, most notably the Second International Mathematics study, researchers had attempted to gather information about factors that influence educational outcomes. That cross-national effort yielded important insights about the effects of several important "opportunity to learn" variables (Anderson, Ryan, & Shapiro, 1989; Burstein, 1993; Walker, Anderson, & Wolf, 1976; Elley, 1992; Elley, 1994; Pelgrum & Plomp, 1993; Postlethwaite, Wiley, Chye, Schmidt, & Wolfe, 1992).

These earlier studies, while informative, raised additional questions about factors singled out as particularly important across the various educational systems (Schmidt, 1992). Not surprisingly, policy makers were eager for more information about these factors than the existing instrumentation and methodology could provide. Critics pointed out that IEA studies did not consider systemic characteristics and their effects in evaluations of achievement and called for more policy-relevant evaluations (see, for example, Theisen, Achola, & Boakari, 1983). This led others to point out the considerable amount of negotiation and compromise required by such international comparative research and also led them to consider ways in which factors, particularly those characterizing system-level considerations, could be better specified in future work (Schwille & Burstein, 1987). Given this, the TIMSS planners decided to include new components for the development and preparation of instruments and research design. Among these new components was the SMSO project and its research program.

Two US funding agencies, the National Science Foundation and the National Center for Education Statistics, awarded a grant in 1990 to form a team of researchers and conduct the SMSO project. SMSO was to perform research in a subset of countries planning TIMSS participation to develop and validate a comprehensive battery of survey instruments addressing student, teacher and school factors that would help understand and explain cross–national student achievement differences in mathematics and the sciences. To prepare for a cross-national study, this research and development project had to be cross-national as well, but on a smaller scale that allowed for more intense investigation. Representatives from several countries were invited to become part of the project. Eventually members of the educational research community from France, Japan, Norway, Spain, and Switzerland joined those in the United States to conduct the various aspects of the project.

HOW SMSO OPERATED

The SMSO was a developmental comparative, cross–national research project. It involved individuals from a number of countries with academic and professional backgrounds in varied disciplines. The research team included those with background and expertise in classroom instruction of mathematics and the sciences, as well as in research methodology, policy analysis and educational psychology. Further, individuals were included who were experienced with quantitative research approaches and others experienced in qualitative approaches. Many in both groups had applied these approaches in previous classroom research.

Members of the SMSO research team worked to develop shared understandings, meanings and conceptual frameworks through discussions of terms, concepts, and research methods. From the beginning, the group attempted to develop a truly interdisciplinary, multi-approach and multinational perspective. Extensive pilot data was gathered as part of the project's instrument validation phase and was analyzed using more traditional quantitative techniques. At the same time the group gathered and drew heavily on a large, complementary qualitative data base, which was investigated using non-traditional, discourse methodology (cf., Goetz & LeCompte, 1984). Researchers from the six countries brought a wealth of disciplinary background knowledge to the project as well. The group included educational psychologists as well as comparative education, curriculum, research methodology, mathematics and science education, and educational policy specialists.

INITIAL INSTRUMENTATION

The task of compiling a database of potential items for survey instruments began before the group's first international meeting. The U.S. National Center for Education Statistics (NCES) funded a series of focus groups to begin identifying issues and appropriate items for specific instruments. These were grouped by questionnaires to be addressed to different members of the educational enterprise — students, teachers, school officials, etc. The development of each questionnaire began with initial conceptual frameworks and models of explanatory factors related to teachers, schools and students. These models were based on existing research literature and on previous IEA studies (e.g., Porter, 1991; Putnam, 1991; Schmidt, Putnam, & Prawat, 1991). The development plan was to put together prototype questionnaires and have SMSO representatives informally pilot these with teachers to obtain their evaluations, insights and suggestions. These early concepts and instruments were largely based on US research literature and on the research background and experience of US representatives.

These prototype instruments were administered to teachers in several countries at both the primary and lower secondary level. The teachers were interviewed to obtain their perceptions and reactions. Some teachers were also asked to maintain detailed logs of the topics, presentation, time allotment, and exercises given students for their mathematics and science classes. The data from these detailed logs were to assist in identifying key research issues and instrument development needs. Both the initial compilation of items and the literature review was dominated by a US perspective. It was thus especially important to verify that these issues and themes could legitimately be found in classrooms in other countries and were important to be considered in a cross-national comparative investigation.

A small number of classroom observations were also conducted in a few countries as part of this initial piloting process. The areas of emphasis in the observations and the teacher questionnaire included the lesson's topic (content), the structure and coherence of the lesson instruction, student grouping patterns during the lesson, instructional and learning activities, interaction patterns of teacher and students, classroom discourse topics and patterns, student participation, and the nature and use of ongoing student evaluation or assessment during the course of the lesson. One focus of the first meeting of international SMSO representatives was the presentation and discussion of teacher interviews and classroom observations for the purpose of evaluating how well the models had been implemented in the instruments.

During this first meeting the discussion among individuals representing various disciplines and countries was complex and almost impossible to track. The creation of cross-national instruments for TIMSS seemed parallel to constructing a "tower of Babel" amidst a confusion of cultural, national and linguistic differences. What became the primary methodology for the project emerged from this apparent chaos. Data sources — interviews with teachers, teachers' responses to prototype questionnaires, summaries of teachers' daily lesson logs, or presentations of classroom observations — were important. The real work of the project became establishing interdisciplinary, multinational dialogue which provided the basis for resolving particular issues as they emerged. This dialogue continued beyond the data to considering the implications the data held for the conceptual foci and possible item wordings for further survey instrument development.

The goal was to produce survey instruments that appropriately and meaningfully assessed key factors in classroom pedagogy which influence students' mathematics and science achievement. This goal remained in the forefront and focused discussions and analyses. Translating the instruments from English into other languages was an issue, but one initially considered more procedural than conceptual or substantive. Many team members had begun with the assumption that all that was needed was refinement of existing prototypes to clarify items for use internationally. They envisioned, at most, small modifications to sharpen prototype items for the present purpose. However, those who had held such assumptions were either soundly disabused of them or, at least, willing to concede that they were problematic by the end of the first meeting.

CHALLENGING ASSUMPTIONS

Representatives of non-US countries arrived at that first meeting with growing perplexity. Teachers in their countries had commented repeatedly that the items on the prototype questionnaires either did not make sense to them or did not apply in their situations. During the first meeting initial presentations were given from country representatives about science and mathematics teaching in their country. This was intended to provide an overview and context for further discussions. Presentations planned to take 15 or 20 minutes took hours — hours usefully spent in listening, questioning and interacting.

English was the common language used for the meetings. Even so, it soon became apparent that the same "language" was not being spoken by all. The problem was not simply one of translation from French or Norwegian or of limited English skills by non-US team members. It became clear that operational definitions for common educational expressions differed among countries. It also became apparent that much was not in common among the participating countries in thinking about education. Even the behaviors and events thought important in characterizing instructional practices were found to differ among the countries. The discussion with participants from other countries began to force the US team responsible for the initial instruments to question their assumptions about the teaching and learning of mathematics and science — assumptions embedded in the prototype instruments.

One example concerns the concept of a lesson. In the United States, the Carnegie Unit was defined in 1906 to standardize amounts of instruction and to facilitate the evaluation of educational experiences. One consequence of this unit conception is that, in the US, the concept of a "period" refers to a regular and specific frequency and to the length of a lesson (Tyack & Tobin,

1994). In the US, a lesson period is typically 50 to 55 minutes five times a week. This concept is so common in the US that it pervaded the questionnaires and US representatives talked about lesson scheduling and sequencing with this implicit underlying conception unaware of the pervasive assumptions made. However, this concept of a lesson was simply not appropriate in many other countries. School days in some of the other countries represented were much more flexible in terms of lesson duration, frequency and sequence of lessons across the school week and throughout the school year.

Two lessons presented at this meeting illustrate the greater variety cross-nationally. One Norwegian science lesson for 13-year-old students lasted for 45 minutes and was one of three taught during the week. The time that the class would meet could change from day to day. One French lesson, also involving 13-year-olds, met for a total of 80 minutes at one time for an extended laboratory or practical experiment. These issues may strike one as purely administrative, with little impact on the substance and development of a lesson. However, the length and frequency of lessons has serious implications for weaving together curriculum and pedagogy — for the manner in which topics are developed within and across lessons as well as for the mode of student participation and activity in any single lesson.

A similar difficulty involved the term "seatwork". "Seatwork" is a commonplace activity in many US classrooms — including science and especially mathematics classrooms. The US concept of "seatwork" again pervaded the prototype questionnaires and was an unquestioned assumption of what occurs in mathematics and science classrooms. Inherent in the US sense of "seatwork" is the idea that students work individually and independently on paper and pencil tasks during the lesson with a minimum of teacher supervision. That is, students are expected to complete tasks without additional information or guidance from the teacher, even though some guidance is occasionally provided. "Seatwork" is a time of independent, individual practice or review.

Students working alone on a task were evident in classroom observations from countries other than the US. However, whether the US concept of "seatwork" applied to these activities needed clarification. The nature of "seatwork" depended on how teachers incorporated this type of student activity into the lesson. The variety with which this activity was used made consistent use of "seatwork" an undefined term that concealed more than it revealed about cross-national differences in science and mathematics classrooms.

In some lessons, as teachers developed a topic, they paused and assigned students a brief exercise or problem to work on immediately. Teachers would give students enough time to think through the problem, if not complete it, and then continue with their presentation incorporating any insights or difficulties students may have encountered with the problem into the teacher's topic development. Clearly students worked independently at their seats with no immediate guidance, but this activity does not fit the U.S. standard "seatwork" concept. This type of seatwork activity involves and motivates students during lesson development, serving as a kind of advance organizer for content yet to be presented. It also provides teachers with some initial information about student understanding.

In other lessons more consistent with the traditional U.S. conception of seatwork, teachers would complete their development of a topic and then assign tasks for students to do for the remainder of the lesson. Differences in how students' "seatwork" was incorporated into the lesson, the amount of time given over to independent student tasks, the nature of the tasks given students, and the manner in which teachers responded to students' tasks or incorporated them into further lesson development, gave lessons a qualitatively different character — even when all appeared to involve at least some "seatwork".

As the above suggests, the prototype questionnaires developed within a US framework presented numerous challenges to non-US representatives. Similarly, descriptions from non-US representatives of teaching and learning in their own country provided some surprising challenges for those from the US. The length of association between a teacher and a group of students is one example. In the US, primary teachers typically teach a single grade and have a different group of students each year. This is not typical in Norway nor Japan. In Norway, primary teachers typically stay with the same group of students, moving with them throughout most, if not all, of their primary school career. The same teacher may be with the same group of students for as many as six years. In Japan, the primary education level consists of three two-year cycles (i.e., lower and upper primary levels). Teachers typically stay with the same group of students throughout one of these two year cycles. These longer patterns of teacher-student association provided a different perspective on some issues, such as teacher preparation time and teachers' understanding of students' knowledge, and, in so doing, revealed subtle assumptions about such issues and how they might relate to other factors of interest.

CHANGING CONCEPTIONS AND METHODOLOGY

Dialogue among country representatives was essential in understanding both reports of classroom observations and teachers' comments about the original, prototype instruments. Each representative came with a way of viewing classroom events that brought certain instructional practices in his or her country to the foreground while leaving others unconsidered. Lying behind or, perhaps, helping to structure these various ways of looking at classrooms were a host of implicit assumptions. These implicit assumptions made it difficult to consider alternatives to typical national practices. The group soon discovered that the narrowing influence of a country "lens" is best overcome through a process of interrogation — by representatives from other countries questioning the commonplace practices within a target country, using their own understandings of educational practice as a basis for inquiry. With implicit assumptions differing among country representatives, the implicit and unexamined soon was made explicit and became the object of group reflection.

The interrogation practice is not aimed at developing "culture free" survey instruments. Such things do not exist. By making explicit some of the implicit assumptions about modal practice within various countries, one can guard against instruments that introduce unwanted, inconsequential variance. This helps ensure that the instruments capture practice differences most likely to have explanatory significance. For example, by clarifying the role of brief exercises in lesson development, one is able to build certain distinctions into instrument items relevant to this practice — and national variations of it — thus capturing its uniqueness compared to brief exercises in other countries. Thus the goal is to construct instruments that reflect the full cross-national complexity of instructional practice and focus on factors most likely to reveal educationally significant differences among typical national practices.

Prototype questionnaire development initially proceeded according to a traditional design-trial-refine model. However, this first meeting made it obvious that the data generated by this approach did not capture the heart of instructional practices across even the six countries. Large cross-national differences were uncovered through the meeting's dialogue on classroom observation reports. At the same time, the information gained from trial responses to the prototype questionnaires did not reveal similarly noticeable cross-national differences.

By the end of that first meeting, the research team concluded that it was necessary (1) to change the methodological paradigm, and (2) to go back into classrooms in the six countries. The prototype instruments simply did not yield a satisfactory picture of the instructional practices in the participating countries. Despite similarity in the surface features of mathematics and science lessons (e.g., Stigler & Perry, 1988), important differences existed among the six countries. A much better understanding of these differences was needed creating instruments meaningful when used for cross-national comparisons.

This initial cross-national dialogue prompted a major reorientation in conceptualization and methodology. The research team concluded that the focus for cross–national studies of contextual factors behind student achievement should not be primarily on concepts that differ quantitatively or distributionally across countries, but rather on concepts that exhibit significant qualitative or categorical differences across countries. The prototypes had been developed with two key assumptions. First, the prototypes assumed implicitly that certain common instructional practices would be found cross-nationally. Second, they assumed implicitly that differences in the frequency and timing of these practices would relate to differences in student achievement. After the first meeting, the notion of "common" practices itself was suspect. Practices are embedded in complex task environments and thus vary qualitatively depending on many factors — including views of teaching and learning, articulation with other practices within and across lessons, how subject matter is defined, and others.

This major re-conceptualization implied developing a new methodology for creating and validating survey instruments. A protocol for conducting comparable classroom observations was developed and a new, more extensive round of classroom observations and teacher log completion was initiated. These new activities were intended to provide data to aid an attempt to more saliently capture in the survey instruments the characteristic pedagogical activities occurring cross-nationally. This shift in conceptualization and methodology started an iterative interaction between qualitative observations and more quantitative survey results from previous studies, and from the prototype pilots aimed at developing useful research instruments and procedures for TIMSS.

Over the course of several years, representatives from the six SMSO countries conducted 127 observations of science and mathematics lessons in the six countries. Professional educators and researchers who were natives of the respective countries conducted the observations and maintained detailed notes. Qualitative summary reports were written by the observers from these detailed notes.

ANALYSES OF CLASSROOM OBSERVATIONS

One of the earliest SMSO efforts was developing common frameworks for mathematics and science. These frameworks were first reported in technical reports (Survey of Mathematics and Science Opportunities (SMSO), 1992a; 1992b) and later reproduced and documented (Robitaille, Schmidt, Raizen, McKnight, Britton, & Nicol, 1993). These provided a multifaceted common language system for examining three aspects of curricular content– the specific topics addressed, what type of student performance may be expected relative to any given topic, and the perspective on the topic students may be encouraged to develop[1]. The frameworks allow meaningful comparisons of curricular content across nations in investigations of curricula, classroom activities and so on. An analytic-empirical method was used to generate these frameworks. Teachers' log data and sample textbooks played an important role in the "empirical" aspect of this development.

These frameworks, combined with some initial items from the prototype questionnaire on instructional strategies and activities, provided the basis for articulating the data gathered from the classroom observations conducted in the six countries. In addition, each country representative was responsible for creating a case study that captured, in as rich detail as possible, standard or typical practice in the teaching of mathematics and science in his or her nation. (Part II of this volume presents edited condensations of these generalized case studies for each of the six countries.) Summaries of specific classroom observations together with the broader information contained in the case studies allowed an image of modal practice in each of the countries to emerge.

One week-long meeting of the international working group was devoted to extensive examination and discussion of the 127 observation summaries and the case studies. Each representative read and analyzed all observations from all countries. After the individual reading and analysis, observations were analyzed both in small groups focused upon specific subjects and age levels (e.g., nine-year-old students' science or thirteen-year-old students' mathematics), and in a meeting involving all representatives. Questions were raised and issues clarified in this latter meeting. The total set of observations served as a common "text" for this process. The discussion process involved lively and at times heated exchanges. The process aimed at developing a shared understanding of each country's modal practices in mathematics and science instruction.

---

[1]Appendix A contains an explanation of each of the three aspects of the mathematics and science frameworks. It also explains and provides examples of the frameworks' hierarchical organization.

Each representative had the opportunity to present her or his own ideas and insights on the observations from other countries. Each non-native representative found some aspects of another country's observations familiar and other aspects more of a surprise, challenging assumptions and expectations. Such instances of surprise and challenge generated questions and the request for further information. The discussion was fruitful and mutually enlightening. In this way each country's observations were analyzed from multiple, often conflicting perspectives. Despite this divergence of perspective, "outsiders" often agreed on what was interesting and surprising about another country's instructional practice. Non-native representatives from different countries frequently had questions or comments on the same aspects of a country's observations.

More importantly, this process of interrogating and analyzing data brought the group closer to a common understanding about key aspects of each country's pedagogy. Based on qualitative data from observations and case studies, and using extended interaction among team members, the understandings emerged in a richer manner than had been possible in previous studies using more traditional quantitatively-oriented instruments and methods.

Near the conclusion of this week-long meeting a list of preliminary conclusions was drawn up for each country. A representative from another country was then given the task of digging deeper into the target country's qualitative data either to confirm or deny the validity of the conjectures about the country's practice. These lead investigators prepared a detailed report of their conclusions. At later meetings, these reports were presented and the original data re-examined. The purpose of this process was to arrive at a characterization of each country's mathematics and science teaching that all participants could support — including the representative from each country involved.

This analysis of classroom observations and case study data across the six countries yielded immediate payoffs. The survey questionnaires were substantially modified to increase their responsivity to pedagogical practice whose importance for cross-national comparisons was revealed by the qualitative data. Further, having achieved this original goal, the research group judged the process so fruitful that they decided to continue the study. This decision was motivated by a sense that there was still more to learn from one another about each country's typical practices in mathematics and science education, and that there was an obligation to share the results of this collaborative effort. The results of these efforts, the insights into the classroom observations and the country level descriptions, are presented in Chapter Three of this volume.

VALUE OF INTERNATIONAL DISCOURSE

The greatest portion of SMSO activity occurred within the six countries as participants gathered data, interviewed teachers, and administered prototype survey instruments. However, the major conceptual work occurred when all the representatives gathered for international meetings. In these meetings the results of the different data collection activities were presented for analysis, discussion and further direction. It was here that new insights, shared conceptual frameworks and a new shared language emerged.

At first the need to clarify perspectives seemed a burdensome preliminary to a more important task. Clarification involved extended, animated discourse among members of the international group, for most in a language other than their own. Making progress on the main task — reaching agreement on instrument items and constructs— was soon clearly seen to depend on the group processes of seeking identification and understanding of the key underlying issues.

The discourse aspect of the multinational meetings took on greater importance over time. Struggling together towards mutual understandings, representatives came to value the "fits and starts", the seemingly insurmountable conceptual obstructions and the rare smooth progress associated with discourse-oriented issue clarification. In particular, the "fits and starts" came to be viewed as signals that further clarification was necessary. Confusion or surprise came to be viewed as essential for clarifying mutual understandings. This discourse process, we believe, has led to insights not possible with a more traditional model — for which development work is done in one country and then validated in others. Working collaboratively with a multinational team from the very beginning in our experience has great potential for yielding instruments that capture the richness and diversity of cross-national educational practice and for revealing salient indicators of difference. Such a team approach, however, cannot be *pro forma* but rather requires time for true mutual discourse, discovery and an interactive style to develop.

Why make such a large point out of what may seem obvious? Most cross-national survey research, especially large-scale studies and particularly in education, is not collaborative in the sense discussed. Researchers from one country usually lead in instrument development and data analysis while researchers from other countries are asked for input and to react to work done in the lead country. This traditional approach to instrument development may be more time-efficient but it lacks the power and insightfulness evident in a discourse-oriented approach such as that utilized for SMSO.

A concrete illustration of the advantages of a discourse-oriented method is in order. An issue debated early by the international team concerned appropriate ways of gathering data on a school's curricular content goals. Some suggested that this information was best gathered from principals or headmasters. The representative from Switzerland had been silently listening to this exchange. After a time he said, "Excuse me, but what is a school and who will fill out the school questionnaire?" The answer to this question seemed so self-evident to most that it was difficult to formulate a response.

The lively discussion that ensued soon made it apparent that the typical practice of obtaining a principal's response to items on a school questionnaire was problematic in Switzerland. Many Swiss schools, especially at the lower level, do not have principals or other administrative staff. Swiss schools are organized and administered differently than those of the other five SMSO countries. This surprise occasioned a further exploration into Swiss practice. What constitutes a school and a school staff emerged as important issues in attempting to understand the Swiss situation (Moser, 1993).

THE CONCEPTUAL FRAMEWORK

SMSO activity was primarily directed at developing context questionnaires for TIMSS survey instruments to assess key factors influencing students' mathematics and science learning cross-nationally. The main focus of this book is on insights and conclusions reached in analyzing the 127 classroom observations and on reflections of the methodology that emerged. However, all of this must be viewed in the broader TIMSS context.

A complete list of products resulting from SMSO activity is given in Appendix B. Among the more noteworthy are the Curriculum Frameworks for Mathematics and Science (SMSO, 1992a and 1992b: Robitaille et al., 1993) and the approach for the TIMSS curriculum analysis, including the associated methodology for analyzing textbooks and curriculum guides (Schmidt & McKnight, 1995; McKnight, Britton, Valverde, & Schmidt, 1992a; McKnight, Britton, Valverde, & Schmidt, 1992b; McKnight & Britton, 1992; SMSO, 1993b). Also important were the student, school, and teacher context questionnaires for assessing relevant factors, issues, and practices necessary to understand differences in student achievement between and within countries (SMSO, 1993c). "Blueprints" for item selection to create achievement tests were still another important SMSO product. The blueprints used for the TIMSS tests were based on a preliminary analysis of curriculum data from selected countries (SMSO, 1993a).

International comparative studies of achievement have often been criticized on methodological grounds. Specific issues include comparability in the definition of student populations and differences in sampling response rates (Rotberg, 1990; Theisen, Achola, & Boakari, 1983). However, the greatest difficulty in cross-national comparisons is beyond the bounds of these more narrow issues. The greatest difficulty in making cross-national achievement comparisons stems from the great variation in educational systems. Other than a rough comparability of function, educational systems — their substance, organization, and mode of operation — display diversity similar to the larger cultural and linguistic contexts in which they are embedded.

The cross-national achievement comparisons, no matter how comparable the achievement measures, must be viewed in context. Results that do not attempt to take into account the nature and quality of the educational experience provided to students are inadequate at best and misleading at worst (Burstein, Oakes, & Guiton, 1992; Schmidt & McKnight, 1995). From the beginning, SMSO project members kept this important consideration in mind. The need to consider educational outcomes in relation to educational inputs has long been a feature of IEA studies. The IEA working model, as explicated in the Second International Mathematics Study (SIMS), assumes various factors influence the educational process at three different levels — system, classroom, and student. These three levels are represented by three "faces" or conceptions of curriculum — the intended, implemented, and attained (see Figure 1-1).

Each of these curricular versions represents a particular set of variables and a societal context in which they are embedded. The intended curriculum refers to the educational system's goals and means. The locus of decision making concerning these factors can vary from local or regional authorities to a central, national authority. Official curriculum visions, aims and goals are presented in national or regional curriculum guides and other documents intended to guide the educational process.

The implemented curriculum refers to practices, activities and institutional arrangements within the educational context of schools and classrooms. These practices occur in implementing the specified visions, aims and goals contained in the intended curriculum. Implementation is influenced by such factors as how subject matter is apportioned across students, classrooms, and lessons; and the background, ideas, attitudes, and pedagogical orientation and practices of teachers.

The attained curriculum refers to the outcomes of schooling — what students have actually attained through their educational experiences. What students learn is influenced by what the system has intended they learn, as well as by the quality and manner in which those intentions have been implemented. Thus, comparisons of educational outcomes — what students have attained — need to include pertinent student background variables, as well as relevant indicators of the system's intended and implemented curricula.

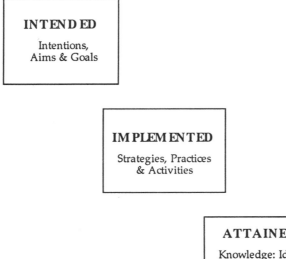

Figure 1-1. Dimensions of Curriculum

Operating from this foundation, the SMSO group undertook the task of creating a fully developed model of educational experience. The model of school level factors was informed by an indicator model of school processes (Porter, 1991). The concepts included in the student questionnaire were informed by a view of student learning influenced by psychological theories of individual differences and by motivational and sociological concepts such as family background. The development of the teacher questionnaire was guided by general psychological and social views of classroom learning consistent with much of current cognitive psychology and influenced to some extent by constructivist education literature in the United States (Putnam, 1991; SMSO, 1993c; Schmidt, Putnam, & Prawat, 1991).

In one of the earliest international SMSO meetings, considerable time and effort was spent exploring the value of various conceptual models that might inform the project's work. The goal was to develop a single model — comprehensive enough to capture commonalities in practice but flexible enough to accommodate differences. This goal was motivated by a recognition of the advantage instrument developers have in being able to refer to a common model that serves as an overall integrating system as they work across sub-domains and sub-systems. Model-driven investigation is fairly common in psychological research, but is used less often in large-scale survey work. More typical of survey studies is having some general idea of what is wanted prior to instrument development but "fleshing out" a model after the fact as a part of the analysis and interpretation of results.

Elements of a more integrated model of students' educational experiences began to be clarified through the discussion process. Four key elements emerged from these dialogues. First is the notion that students' curricular experience reflects the complexity of the educational system as a whole. Many factors have an impact on education, even at the classroom and student level, and all are systemic — impacts of the broader context of an educational system. What occurs is a function of the entire system. Efforts to identify the effects of a single, isolated aspect of the system fail because of the interrelated nature of educational systems.

A second important notion is that any given system's provision of possible educational experiences is limited — no system can provide all possible experiences. Decisions made have implications not only for what will be included in these experiences but also for what will not be included. Students' school time is a finite resource. Attempting to "do it all" and include all possible learning experiences ignores this limit and, substituting for the wise allocation of time resources, can have disastrous consequences for the whole educational enterprise.

A third notion is that curriculum has both an intended and implemented side. The former is found in teachers' guides, textbooks, and system-wide curricular policy documents. The latter is found in the selection of topics and pedagogical approaches, and in the provision of educational experiences to develop topics.

The above suggests that subject matter is a key aspect of classroom educational experience. As the opening scenarios illustrate, content is at the heart of schooling. The teacher's role is pivotal. Teachers serve as the final arbiters of curricular intentions and they are the "brokers" or "midwives" of students' content-related learning experiences (Schmidt & McKnight, 1995; Contreras, 1990; Schwille, Porter, Belli, Floden, Freeman, Knappen, et al., 1983).

A workable model describing delivery of content-related educational experience was refined at one international SMSO meeting. This model was modified and elaborated in subsequent international meetings of the SMSO group. An overview of the model is presented in Figure 1-2. It is an expanded version

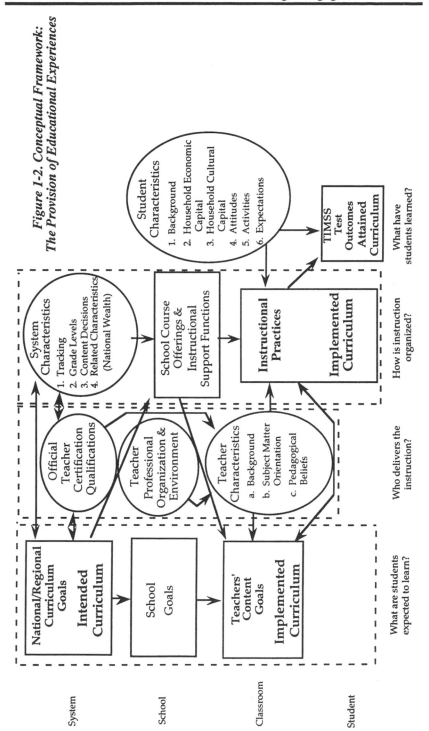

Figure 1-2. Conceptual Framework:
The Provision of Educational Experiences

of the simpler curriculum model in Figure 1-1. (The three key elements from Figure 1-1 are included in bold print in Figure 1-2). The constructs labeled "Intended," "Implemented," and "Attained" are fitted into a column and row structure. The columns represent four key research questions (e.g., "What are students expected to learn?"). Each of the four questions is fundamental to cross-national investigations. The rows represent different levels of a single overall educational system. The arrows in the figure indicate the network of relationships among the constructs. The model, while difficult to assimilate at once, is a useful device for examining a potential cross-national comparison of educational systems and for identifying others that might be made. The questions at the bottom of each column are useful organizers for further elaborating the model. (see Figure 1-2)

What are students expected to learn? Describing students' possible learning experiences begins with describing what knowledge and skills students are expected to attain. There are three main levels of the educational system at which such goals and expectations are commonly set: the national or regional level, the school-site level, and the classroom level. The first question involves not only determining learning goals for a system or country as a whole, but also differentiating such goals for sub-divisions within the larger system — regions, tracks, schools, grade levels and so on. Learning goals specified at the national or regional level are, in the traditional language of the IEA, the intended curriculum. Learning goals specified at the school or classroom level are part of the implemented curriculum. Curriculum, as goals and as planned distribution of educational opportunities, is addressed at each of the four modeled levels of educational systems.

Who delivers the instruction? Students' learning experiences require instructional activities and are undeniably molded by the teachers who deliver instruction. Factors influencing the teacher's key role can be investigated by examining official teacher certification qualifications — including grade and subject restrictions, educational attainment requirements for awarding each license, type of educational degree required, and any specific course work or practical experiences that may be additionally required. The professional organization and environment in which teachers work also influences their instructional efforts. This includes time usage — that is, the proportion of professional, school-day time spent in planning, the proportion of time teaching mathematics or science, and the amount of cross-grade level teaching in mathematics and the sciences (Doyle, 1986; Lockheed, 1987), and so on. Cooperation and collaboration among teachers in planning instructional sequences and strategies also has an impact on what occurs within classrooms.

Teacher characteristics play a role, influencing the quality of instruction and thus the quality of students' educational experiences. These include such factors as teachers' backgrounds and teachers' beliefs (see Porter, 1991). Teacher background variables include age, gender, education, subject taught, and teaching experience. Teacher beliefs address subject-matter orientation and subject-matter specific pedagogy. Teachers' beliefs about subject matter can affect instructional practices and student achievement (Thompson, 1992; Putnam, 1992; Peterson, 1990). These subject-matter beliefs also include the views a teacher has of mathematics and the sciences as disciplines. Pedagogical beliefs, by contrast, deal with teachers' beliefs about good ways to teach particular topics in mathematics or the sciences.

How is the instruction organized? The organization of instruction influences the implemented curriculum and students' learning experiences. Decision making concerning instruction is widely distributed — including the very top of the educational hierarchy, intermediate levels, and school administrators, as well as classroom teachers. Major organizational aspects include variations in the age-grade structure of educational systems, the nature of the schools that serve different arrays of grades, and the various curricular tracks or streams into which students are placed. Economic resources also influence how the instruction is organized, as well as influencing the qualifications of the teaching force, the array of instructional resources available to those teachers, and the time and material resources available for students.

Instructional organization also subsumes school course offerings and support functions for mathematics and science instruction. The organized implementation of curriculum in classrooms is also involved, including the role of students in classrooms (the reason the box representing this concept extends across both the classroom and student levels in the model) and factors related to it — textbook usage, lesson structure, instructional materials, assessment of students, teacher and student interaction, homework, and in-class grouping of students (Ramseier, Moser, Reusser, Labudde, Buff, 1994; Moser, 1995). These factors are represented in a sub-model shown in Figure 1-3 — which expands the "Implemented Curriculum" element in Figure 1-2. The implemented curriculum model in Figure 1-3 includes both "Teacher Content Goals" and "Instructional Practices". It also suggests the relationship of subject matter orientation and pedagogical beliefs to what teachers do (Salomon, 1992; EDK, 1995; Stebler, Reusser, Pauli, 1994).

What have students learned? Investigating learner characteristics is an important part of understanding what and how students learn. Beyond the influence of curriculum goals, teachers, and instructional organization, student characteristics influence what is learned from particular educational experiences provided. These characteristics include students' academic histories, economic capital of families (socio-economic status), cultural capital of families, students' self-concepts, how students spend time outside of school, students' beliefs, and students' motivations, efforts and interests pertaining to school and particular subject matters.

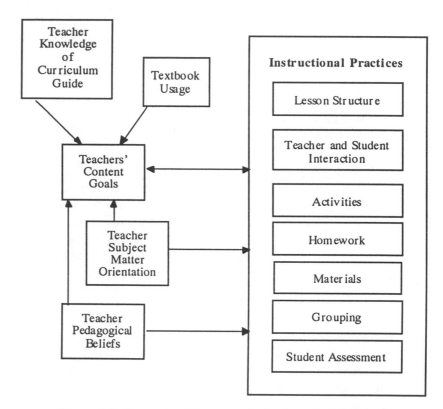

Figure 1-3. Conceptual Framework: Implemented Curriculum

It is neither possible nor desirable to identify and measure every possible factor that affects an educational system — or even all of those portrayed in the model here. However, this model of students' educational experiences recognizes the interconnections between major components of the educational system in a way analogous to conceptualizations of many proponents of systemic educational reform (Coll, 1987; Gimeno & Pérez Gómez, 1992; O'Day & Smith, 1993; Clune, 1993). This is a generic model useful for describing many specific educational systems. It does not advocate a particular system but rather is intended as a template against which to identify systemic variations. In this sense it is particularly useful for cross-national comparisons of educational systems.

CONCLUSIONS AND IMPLICATIONS

Traditionally, researchers who conduct small-scale, relatively circum-scribed cross-national studies have worked independently of those who conduct large-scale studies like TIMSS. The former take pride in their eclectic approach, at times mixing qualitative and quantitative research methods to do whatever is necessary to understand and illuminate specific practices in two or more cross-national sites. Their colleagues conducting large-scale research typically have less maneuverability in choosing a blend of quantitative and qualitative methods. Quantitative methodologies often seem to be their only practical methodological option.

The SMSO project tried to marry these two approaches, combining dis-course-driven qualitative methodology with traditional quantitative instrument validation methods to develop more sensitive and responsive questionnaire instruments. This hybrid approach is more labor intensive than traditional survey methodology, but the SMSO experience suggests its merits are enough to warrant careful consideration by others engaged in similar enterprises. There are, however, some caveats. The discourse-driven approach to instrument development necessarily involves a small number of individuals (from more than one country). The time, effort, and personal energy required to arrive at shared understandings cannot emerge in the context of large meetings operating with more formal modes of communication and organization. These required investments are not guaranteed by gathering international groups, large or small. The requirements of the approach must be considered — diversity in representation, flexible time schedules, organizational arrangements that facilitate productive discourse, and so on. The SMSO project illustrates how a small group can capitalize on its size and diversity to investigate problems relevant to a large-scale, cross-national research study.

The innovative products growing out of the SMSO project were many: (1) a conceptual "systems model" of students' educational experiences to guide cross-national investigations; (2) new research methodologies (those for the curriculum analysis); (3) a new way to link achievement test specifications to subject matter specific curriculum analyses and questionnaires through the use of a common language system (i.e., the TIMSS curriculum frameworks); (4) an enhanced national–context sensitive way to develop and validate survey instruments; and (5) a new way of looking at what goes on in classrooms.

The SMSO outcomes itemized above illustrate the benefits that can be derived from mixing large- and small-scale approaches, and from blending qualitative and quantitative methodologies. The smaller, more in-depth activities can serve as a resource or "creative workshop" for the large-scale, cross-national study. To accomplish this goal, however, a relatively small group of researchers from several countries must invest the time and energy necessary to achieve shared understandings about important aspects of education. Such a commitment was sustained by the SMSO research team. More than anything else, this commitment (the need for which was originally unanticipated) contributed to the SMSO project's successes.

Chapter 2
Exploring the Story of Curriculum: Examining Artifacts of Intention

A Norwegian science teacher begins her eighth grade class by asking students what results they obtained in the laboratory exercise they completed the day before. The exercise involved measuring the temperature of ice as it melted and the temperature of water as it was heated to the boiling point. The teacher asks students for reasons explaining their results.

A student draws a temperature graph on the board. Together the teacher and her students discuss the shape of the graph around the melting and boiling points and possible sources of error. Other students draw similar graphs using their own previously collected data. The teacher concludes this part of the lesson by explaining the relationship between temperature and energy as illustrated in the lab exercise.

The teacher then introduces a related topic — the water cycle. As she talks, she asks students questions about evaporation, water resources, and rain. They respond to her questions. A few students mention other aspects of the water cycle such as climatic changes and the water table. The lecture and discussion occupy the final 20 minutes of class time.

In Japan, a teacher begins his eighth grade science class by reminding students they have just completed lessons on atmospheric pressure. He tells them they will now study something about water and he tells them where this material is found in their textbooks. He continues by writing a question on the blackboard, "Where does the water go after a rainfall?" Students copy the lesson topic and the question into their notebooks. The teacher asks several related questions and then tells students to summarize their ideas in their notebooks.

After a few minutes, the teacher has several students share their ideas with the class. The teacher poses more questions such as "What happens to the water in wet laundry?" The teacher moderates a discussion about this question. In the discussion students bring up the idea of evaporation. The teacher reminds them that they studied the three states of water in science class the year before. He states that today's topic — relative humidity — is related but not the same. He refers students to a graph of temperature and humidity in their textbook. The teacher discusses the graph with the students.

The teacher then has students use the graph to figure out how much water the air in their classroom could hold at 20º C. Students work on the problem in their notebooks. After students have worked on the problem for awhile, the teacher has one student solve the problem on the blackboard. There is some confusion about the proper measurement units so the teacher explains how to go from the measurement units used in the graph to those needed for the classroom.

The teacher poses a similar problem and again has students work in their notebooks. The teacher comments that he would have a student demonstrate how to do the problem but since many were unable to figure it out, he will explain the problem at the blackboard. The end of class is signaled and the teacher tells students to memorize the equation in their textbook.

CURRICULUM IN THE CLASSROOM

The story of curriculum continually unfolds in classrooms around the world. Teachers and students enact well-defined roles moving towards time-honored ends. The classroom is the center of this curriculum story but it does not begin there. It begins with curriculum's shapers and molders through their selection and adoption of specific visions, aims, and goals to guide instruction. The story continues as specific visions, aims and goals are expressed in particular curricular materials and resources that may support and guide classroom instruction. The present chapter looks at the parts of this story occurring before classroom instruction. It examines specific curricular visions and goals and their products — curriculum frameworks and textbooks. This examination is a partial response to one question posed in the Provision of Educational Experiences model discussed in Chapter One: "What are students expected to learn?" The specific aspect of the model explored here is shown in Figure 2-1.

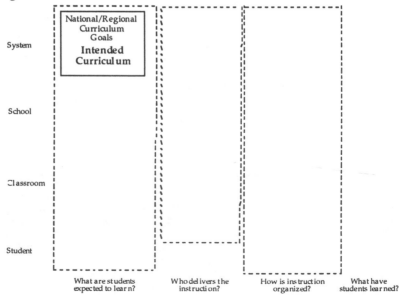

Figure 2-1. The Provision of Educational Experiences
– Intended Curriculum

The remainder of the book, especially the following chapter, explores more fully issues of curriculum as it is enacted in classrooms. That discussion is mostly based on classroom observations. Those observations focused on teachers' instructional practices. Even so, subject matter content played a fundamental role in understanding each observed lesson.

The classroom curriculum story, seen in teacher activities and interactions with students, and the planners' curriculum story, documented in curriculum frameworks[1] and textbooks, are not simply parallel stories unfolding in the same setting. On the contrary, classroom observation indicates that dimensions of curriculum's content interact with aspects of classroom activity to yield specific lessons — lessons qualitatively different from each other. A thorough understanding of our observation's classroom activities required characterizing several aspects of the lesson's content — the topic, how the teacher dealt with the topic, and what students were expected to do with the topic.

It was impossible to discuss typical instructional practices apart from this characterization of relevant subject matter. Instructional practices are subject matter specific, not generic. Classroom activity is thoroughly embedded in and structured by the nature of the discipline involved (Shulman, 1986a; Shulman, 1986b; Stodolsky, 1988; Hernández & Sancho, 1993; Stodolsky & Grossman, 1995). Thus some characterization and understanding of relevant subject matter content is a prerequisite for any analysis of instructional practices.

CURRICULUM COMPLEXITY

Considering various facets of a lesson's subject matter content makes it possible to create a complex, composite portrayal of the curriculum story unfolding in classrooms. A lesson's content is complex. It has many aspects including describing precisely the lesson's topic, how the topic is presented, and what behavior is expected from students interacting with the lesson topic. Each of these three facets may be explored by examining curricular plans and intentions.

---

[1] Curriculum frameworks articulate official policies as they apply to large groups of students; for example, all students in a certain grade or a specific type of school. In short, these set out curricular intentions. Each is an official and global guiding statement for what a curriculum is intended to be in a specific context and how instruction is to be conducted. Such documents can have different titles such as curriculum guides, National Curricula, etc.

Complexity is used here not as a judgment on the utility, value, or academic prestige of any particular curriculum. "Complexity" was chosen to suggest that lesson content is a structured pattern of interconnections, both within and among various aspects of the curriculum's subject matter.

School lesson topics are not isolated, discrete elements in any subject matter domain. They maintain rich conceptual and practical connections to other topics. Perceiving and appreciating topic interconnections — a discipline's "deep structure" —distinguishes the expert from the novice (VanLehn, 1989). Indeed, some lessons' object is discovering, identifying, and making clearer such topic interconnections.

Complexity is used here only descriptively. This chapter's characterizations of country curricula are not evaluations nor do they make recommendations among curricula, either explicitly or implicitly. These characterizations are meant only to describe the rich diversity within and among the different facets of a curriculum's subject matter content.

The first curricular facet explored is topic or conceptual complexity. This simply refers to the basic identification and characterization of curriculum topics. For example, some questions relevant to this aspect of content include: "What topics are addressed in primary grade science?", "Do all primary students study the same science topics?", and "What topics are emphasized in primary grade mathematics?"

This topic complexity aspect also involves considering whether all topics are equally important. Some topics are more basic. Other topics are more difficult — that is, they require students to understand and integrate a larger, more diverse conceptual background. For example, adding two single digit whole numbers is typically considered more basic and less conceptually difficult than multiplying two vectors or matrices. Similarly, building understanding of different types of animals and plants seems less involved than discussing thermodynamics and energy in anything other than the simplest terms.

The second curricular facet explored is developmental complexity. This can refer to at least two different types of topic development or treatment — the way topics or sub-topics are sequenced and developed within a single lesson, and the way topics are sequenced and developed across an entire curriculum. Are the topics addressed in primary grade mathematics included, reviewed, or emphasized in lower secondary mathematics? When do students learn about electricity? How does a curriculum distribute emphasis on this topic across the grade levels?

Different curricula show different topic sequences and arrangements. For example, topics may be presented in clusters and studied in a focused and concentrated manner. Further study may build on or apply this focus but the curriculum moves on to focus on new topics. In other instances, topics may be presented by spiraling —returning again and again in the curriculum to the same topics but seeking new depth each time. Developmental complexity may also refer to the pedagogical presentation of a topic — the logic, coherence, and general manner in which a topic is handled within a lesson. This chapter addresses some questions about the first type of developmental complexity; later chapters deal with the second.

The third aspect of curricular content complexity, cognitive complexity, is not a function of content *per se* but a function of the pedagogical intention for a particular topic. Given any particular topic, there are many things students might be expected to do with that topic's content. Are students expected to memorize definitions? to learn a procedure? to understand a relationship between two concepts? to explain their solution to a problem that has been posed? The possibilities are many and varied.

As with topic complexity, common sense suggests that some student performances require greater thought, reflection, and knowledge than others. Differences in cognitive complexity can only be examined after careful description of what students are actually expected to do with individual content. Student performance expectations are broadly catalogued as one aspect of the TIMSS curriculum frameworks (developed in SMSO technical reports and made more widely available in Robitaille et al., 1993). The mathematics framework's performance expectations, for example, distinguish between using routine procedures (counting, multiplying, measuring, etc.) and using more complex procedures (estimating, organizing and displaying data, comparing and contrasting two representations., etc.).

The science framework similarly distinguishes between understanding simple information (e.g., vocabulary and symbols) and understanding more complex information (how pressure raises the boiling point of liquids, how fire is a part of pine trees' life cycle, etc.).

TRIPARTITE MODEL OF CURRICULUM

Each of these aspects of content — topic, developmental, and cognitive complexity — contributes to a curriculum's character. Each may be examined by studying documents that embody the visions, aims, goals and intentions of a given system's curriculum. Curriculum emerged as a significant explanatory factor in previous IEA studies (McKnight et al., 1987; Schmidt, 1992; Wiley, Schmidt, & Wolfe, 1992). Curriculum has been viewed in three aspects

— the intended, implemented, and attained curriculum. (See Figure 1-1, page 17 and related discussion). Document study is useful in examining the first of these aspects.

Intended curriculum refers to an educational system's goals and plans. Decision making for the intended curriculum takes place at different levels — local authorities, regional authorities, at the national level, and so on. Official curriculum visions, aims, and goals are presented for educational use in national or regional curriculum guides, and in other documents intended to guide and direct the educational process.

Textbooks are another kind of document that embody curriculum visions, aims and goals. However, textbooks may be consistent in varying degrees with documents explicity presenting official visions, aims, and goals. This relates to textbooks' official status and how they are published. In some systems textbooks are official government or educational system publications. In other systems textbooks are produced commercially and compete for official adoption. In yet other systems, textbooks have absolutely no official status and develop totally as commercial projects. Even in this last case, textbooks are likely to reflect, at least in part, official intentions simply for commercial viability.

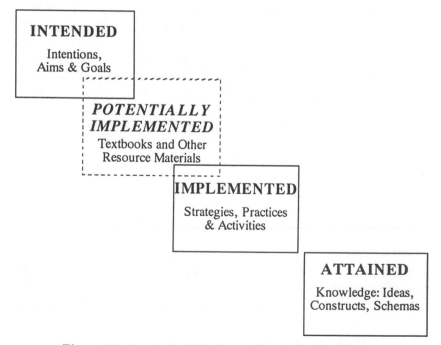

Figure 2-2. Textbooks Relation to Curricular Dimensions

The relationship between textbooks and official intentions contained in curricular frameworks is similarly varied. Textbooks can be entirely consistent with official intentions or, in other cases, provide more of a bridge between what is intended and what is "implemented" in classrooms. Figure 2-2 modifies the IEA curriculum model previously presented to show the unique role of textbooks.

Textbooks embody curriculum intentions in specific pedagogical approaches and resources. Textbook content can, but does not necessarily, help guide what individual teachers do in their classrooms. Textbooks thus represent the potentially implemented curriculum. They move beyond stating aims, visions, and goals by presenting selected curriculum topics in a specific sequence and making use of particular pedagogical approaches — explanations, examples, activities, and exercises. Textbooks embody one set of curriculum goals that can inform, guide, and support classroom instruction. Previous studies have found that textbooks have significant impact on teachers' classroom instruction (D'Ambrosio, 1985; Carraher, Carraher, & Schlieman, 1987; Lockheed, 1987; Schmidt, Porter, Floden, Freeman, & Schwille, 1987; Sosniak & Perlman, 1990; Valverde, 1993). However, the actual impact any text has on how curriculum is implemented in a classroom obviously depends on how a teacher chooses to utilize the textbook as a resource. The individual teacher's use of textbooks itself might be influenced by national curricular policies, regulatory practices, and teacher preparation methods.

CURRICULUM ANALYSIS

Given the theoretical, conceptual and empirical importance of curriculum, the research design for TIMSS included a comprehensive investigation of curricula for mathematics and the sciences. The investigation balances the need for detailed curriculum topic information at target grade levels and a broad characterization of topic introduction and inclusion across all the years of schooling (Britton, 1992; McKnight, 1992; McKnight & Britton, 1992; Robitaille et al., 1993; SMSO, 1993b, 1995f).

The TIMSS curriculum analysis, conducted in each country by educational experts with special training, characterizes curriculum frameworks and textbooks in detail. It uses the TIMSS science and mathematics curriculum frameworks[2] for this and, for selected topics, a description of their sequence and emphasis across the years of schooling (SMSO, 1995a, 1995b, 1995c, 1995d, 1995e, 1995f). This analysis provides a way to describe and characterize an

---

[2] See Appendix A for an explanation of and some examples from the TIMSS frameworks.

educational system's mathematics and science curricula, including aspects of the three types of content complexity discussed above. The curriculum frameworks' content categories (and the numerical codes attached to them) provide a common language for describing and comparing topics across country curricula. The performance categories and codes do the same for describing what students are expected to do with each curriculum topic.

Comparing included topics in country curricula at specific grade levels gives some insight into topic complexity. These data are found for curriculum frameworks and textbooks at each of the TIMSS student population levels.[3] Comparing the number of topics addressed at a specific grade level, and the number of new topics added and old topics dropped at that level, provides insight into the curriculum's developmental complexity. These data are based on general topic trace mapping and in-depth topic trace mapping developed for TIMSS (McKnight, Britton, Valverde, & Schmidt, 1992a; McKnight, Britton, Valverde, & Schmidt, 1992b).

Finally, the TIMSS' frameworks performance expectation codes give some insight into curricular cognitive complexity by detailed description of what students are expected to do with each curriculum topic.

PORTRAITS OF CURRICULUM

This chapter's goal is to paint national "portraits" of the curriculum in mathematics and in the sciences for the TIMSS target grades. These portraits provide a context for understanding the more in-depth treatment of classroom instruction in the next chapter. The curriculum data sources in both chapters focus on the same target populations. This chapter's broad-stroke portraits — based on analysis of curriculum documents — are composite pictures when more than one document of a particular type (e.g., mathematics textbooks) existed for a particular grade level in a particular country.[4]

These broad curricular portraits help "situate" the more focused treatments of curricula in the case studies. They help in assessing how typical were

---

[3] The curriculum frameworks and textbooks analyzed are those dealing with the TIMSS focal grade levels. These are analyzed by partitioning the documents into relatively homogeneous segments and coded according to the three aspects of the relevant TIMSS Curriculum Frameworks. This yields measures such as the percentage of segments addressing a particular topic or performance expectation. These are the percentages presented and discussed in this chapter.

[4] Explanations of the procedures used in various curriculum analyses are found in SMSO technical reports — see SMSO, 1995a, 1995b, 1995c, 1995d, 1995e, 1995f

particular cases' treatments of curricula. They provide some check against unwanted bias in the topics sampled in the lessons observed — a concern with SMSO's qualitative studies with selective sampling of cases and lack of systematic coverage of all possible content topics.

Any portrait, no matter how carefully constructed, is always open to different interpretations. This is particularly true when some expected common features are absent in a case. For example, there is no evidence of teachers using a textbook in our observations of science teaching in Switzerland. Was this a "sampling" accident or was it common practice? The curriculum analysis supports the idea that no official science textbook exists for primary students, at least in the German canton focused on in the SMSO observations.[5]

It is unwarranted to conclude, based on the absence of an official textbook, that science was not taught in the Swiss canton observed. A curriculum guide existed at the primary level and observations revealed that science was taught. For these lessons, teachers used a variety of resource books containing information on plants and animals rather than relying on a single textbook. The absence of textbooks has implications for teachers and students in Swiss primary grades. However, these implications are best grasped by close, detailed study as in the qualitative case studies. The broad-stroke curriculum analysis and the finer-grained case study analysis complement one another. A further look at the broad portraits is needed before moving to the **case studies**.

WHAT ROLE DO CURRICULUM FRAMEWORKS AND TEXT-BOOKS PLAY IN EACH COUNTRY?

The specific role of textbooks and official curriculum frameworks differs among countries depending largely on how centrally administered the educational system is. Curriculum frameworks can be official publications of a centralized government agency or created at a regional or local level. France, Japan, Norway and Spain have national curriculum frameworks detailing curricular goals to be followed in all regions and schools. There is no one national curriculum guide in Switzerland or the United States. In Switzerland, recommended curriculum frameworks are created within each canton. In the United States most states have detailed curriculum frameworks and many districts within states also have curriculum guides.

---

[5] Switzerland has German, French, and Italian cantons —each with their own curriculum frameworks and textbooks. All classroom observations discussed in the next chapter came from one of the German cantons, so only the documents for the German cantons are included in this chapter's analysis.

The role that these frameworks play in classroom accomplishment of curriculum goals also differs. Some frameworks mandate both content topics and pedagogical approaches to be used in classrooms at each educational level. Others are less detailed and considered more as recommendations than mandates. In France, Japan, Norway, Spain, and Switzerland all schools are expected to follow the relevant curriculum guides. The Japanese frameworks also include specific pedagogical examples for some topics. Even so, in each case teachers have considerable freedom in shaping classroom activities to meet these official guidelines and accomplish their goals. In the United States, the extent to which official frameworks actually affect classroom practice is a matter of considerable debate (for example, see Schmidt et al., 1987; Prawat, Remillard, Putnam, & Heaton, 1992; Putnam, et al., 1992).

There is similar diversity in the official status of textbooks within educational systems. In five of the six SMSO countries, textbooks are produced by commercial publishers rather than the government. Mathematics textbooks in Switzerland's German cantons are the single exception. These textbooks are published by the government but consist mainly of student exercises. In France, Japan, Norway, and Spain, the commercial publishers producing textbooks must adhere to detailed national curriculum guidelines. In Norway the government approves textbook selection. In France and Spain, groups of teachers at the school or other levels select textbooks for use.

In the United States, textbook selection is made in one of several ways. Some states either select textbooks for the state as a whole or approve a few textbooks from which local districts or schools may choose. In other states, textbook selection is made at the school district level. In many situations teachers at the school level decide which textbook will be used.

The case studies for each country in Part II of this volume present further information needed for understanding how mathematics and science curricula are reflected in classroom practices in each SMSO country. Even with great diversity in the official status and practical role of curriculum frameworks and textbooks, examining these documents provides some insight into characteristic national versions of school mathematics and science.

WHAT IS INTENDED IN THE MATHEMATICS AND SCIENCE CURRICULA IN THE SIX COUNTRIES?

Discussion returns now to the issue of topic or conceptual complexity. Relevant questions about this aspect of content include, "What are the topics addressed in primary science?", "Do all primary students study the same science topics?", "What topics are emphasized in primary mathematics?" and other similar questions. Table 2-1 presents a summary of topics commonly intended in the curricula of the six countries. This table summarizes analyses of curriculum frameworks and student textbooks for TIMSS Population One and Two[6] in each of the six countries. The table lists only those topics found in documents in at least five of the six countries. A topic is included even if only mentioned in a curriculum guide. Topics mentioned in both curriculum guides and textbooks are presented in the second column.

As Table 2-1 reveals, there was considerable overlap among curriculum frameworks across the six countries in the topics intended for students. The number of intended topics was greater for students at Population Two (usually lower secondary level) than at Population One (the primary level). This was true for both mathematics and science. The number of science topics intended for students increased dramatically at the lower secondary level compared to the earlier grades. This reflects both the great diversity within science and the fact that multiple courses of study focusing upon different branches of science were often offered for students at this level. Dividing intended science topics among three broad branches of science, each science area's number of topics was very similar to the number of commonly intended mathematics topics — mathematics (17) compared to earth science (11), life science (14), and physical science (16).

Common intentions were also seen in cognitive complexity — what students are expected to do with topics — found in curriculum frameworks and textbooks. Table 2-2 presents performance expectations commonly intended by at least five of the six countries. In contrast to the number of commonly intended Population Two (lower secondary) science topics, the number of commonly intended Population Two performance expectations did not differ dramatically from what was found in mathematics. The number of commonly intended Population One (primary) mathematics performance expectations (20) was more than triple the number commonly intended for Population One science (6). This may indicate that, even though the number of Population One mathematics topics (7) was about equal to that of Population One science (10), there was a greater emphasis on students' versatile mastery of mathematics at Population One. That is, there might not only have been more focus in mathematics content (i.e., fewer topics) but more diversity in expected performances as well.

---

[6] TIMSS student populations are defined according to students' age: nine-year-olds are Population one, thirteen-year-olds are Population two. The TIMSS study design includes the two adjacent grade levels containing most students of the appropriate age. The upper of the two adjacent grade levels is considered the TIMSS Target Grade. In this chapter, all references to documents for TIMSS student populations are for the Target Grade.

Table 2-1. Commonly Intended Topics Among Six Countries

Common Topics In Curriculum Frameworks Only	Common Topics In Frameworks & Textbooks
Population 1 Mathematics Topics	
• Measurement: Perimeter, Area & Volume • 3-D Geometry	• Meaning of Whole Numbers • Whole Number Operations • Properties of Whole Number Operations • Measurement Units • 2-D Polygons & Circles
Population 2 Mathematics Topics	
• Rational Numbers & Their Properties • Measurement: Perimeter, Area & Volume • Measurement: Estimation & Errors • 3-D Geometry • Proportionality: Concepts • Proportionality Problems • Uncertainty & Probability	• Meaning of Whole Numbers • Decimal Fractions • Integers & Their Properties • Measurements Units • 2-D Geometry: Basics • 2-D Geometry: Polygons & Circles • Geometric Transformations • Patterns, Relations & Functions • Equations, Inequalities & Formulas • Data Representation & Analysis
Population 1 Science Topics	
• Biomes & Ecosystems • Interdependence of Living Things • Animal Behavior	• Plant & Fungi Types • Animal Types • Microorganism Types • Organs & Tissues • Pollution • Land, Water & Sea Resource Conservation • Material & Energy Resource Conservation
Population 2 Science Topics	
• Land Forms • Earth's Atmosphere • Earth Building & Breaking Processes • Earth In The Solar System • Planets • Beyond the Solar System • Microorganism Types • Cells • Biochemical Processes In Cells • Life Cycles • Evolution, Speciation, & Diversity • Biomes & Ecosystems • Habitats & Niches • Interdependence Of Living Things • Animal Behavior • Macromolecules & Crystals • Subatomic Particles • Energy Types, Sources & Conversions • Magnetism • Explanation of Chemical Changes • Rate Of Chemical Change & Equilibria • Organic & Biochemical Changes • Nuclear Chemistry • Electrochemistry • Nature & Conceptions Of Technology • Influence Of Science & Technology On Society • Material & Energy Resource Conservation	• Bodies of Water • Rocks & Soils • Weather & Climate • Earth Physical Cycles • Earth's History • Animal Types • Organs & Tissues • Organism Energy Handling • Organism Sensing & Responding • Human Diseases • Classification Of Matter • Physical Properties Of Matter • Chemical Properties Of Matter • Atoms, Molecules & Ions • Electricity • Descriptions of Chemical Changes • Energy & Chemical Changes • Pollution • Land, Water & Sea Resource Conservation • Food Production & Storage • Science & Other Disciplines • History of Science & Technology • World Human Population • Effects of Natural Disasters

Table 2-2. Commonly Intended Performance Expectations Among Six Countries

Common Topics in Curriculum Frameworks Only	Common Topics in Frameworks & Textbooks

———— **Population 1 Mathematics Performance Expectations** ————

• Predicting • Verifying • Developing Algorithms • Generalizing • Conjecturing • Justifying & Proving • Axiomatizing • Using Vocabulary & Notation • Relating Representation • Critiquing	• Representing • Recognizing Equivalents • Recalling Math Objects & Properties • Using Equipment • Performing Routine Procedures • Formulating & Clarifying Problems & Situations • Developing Strategy • Solving • Developing Notation & Vocabulary • Describing/Discussing

———— **Population 2 Mathematics Performance Expectations** ————

	• Representing • Recognizing Equivalents • Recalling Math Objects & Properties • Using Equipment • Performing Routine Procedures • Formulating & Clarifying Problems & Situations • Developing Strategy • Solving • Developing Algorithms • Conjecturing • Using Vocabulary & Notation • Relating Representations

———— **Population 1 Science Performance Expectations** ————

• Identifying Questions to Investigate • Designing Investigations • Conducting Investigations • Interpreting Investigational Data	• Understanding Simple Information • Understanding Complex Information

———— **Population 2 Science Performance Expectations** ————

• Abstracting & Deducing Scientific Principles • Identifying Questions to Investigate	• Understanding Simple Information • Understanding Complex Information • Constructing & Using Models • Making Decisions • Using Apparatus, Equipment, Computers • Doing Routine Experimental Operations • Gathering Data • Organizing & Representing Data • Interpreting Data • Designing Investigations • Conducting Investigations • Interpreting Investigational Data • Formulating Conclusions from Data • Accessing & Processing Information • Sharing Information

WHAT DIFFERENCES EXIST IN COMMON INTENTIONS AMONG THE COUNTRIES?

The six countries obviously shared much in their mathematics and science curricula. There was broad agreement on which topics were most appropriate for students at the two age levels examined. There was also considerable agreement on what students are expected to do with these topics. However, as the opening narratives illustrated, the manner in which these commonalties were implemented could have differed markedly.

The commonalities found would hardly be enough to predict the qualitative differences seen in the classroom observations. Topic selection, however, is only one aspect of curriculum complexity — and discussion thus far has focused on its commonalities rather than its diversity. Qualitative differences observed in classrooms might also be related to the emphasis topics were given and how they were developed across the years of schooling (developmental complexity). Qualitative differences observed might also relate to diversity in the expectations held for students in the topics considered (cognitive complexity). Table 2-2 considers only commonality of expectations, not diversity. Variation in these aspects should result in very different curricular portraits even in a context of common intentions.

Textbooks turn intentions into pedagogical resources potentially informing and influencing classroom activities planned to attain curricular aims and goals. Textbooks must embody not only appropriate topics but acceptable pedagogy. They provide a transition from direct expressions of intentions in curriculum frameworks to classroom practices "implementing" those intentions in learning opportunities. Examining how textbooks treat topics sheds further light on similarities, differences, and distinctions in provisions for learning mathematics and the sciences, and on the sources of qualitative differences seen in classroom observations.

Despite the commonality demonstrated across the six countries (see Tables 2-1 and 2-2), there was also considerable diversity. There was variation in all facets — topic, developmental, and cognitive complexity. Topic complexity variation was apparent both in emphasis given commonly intended topics and in emphasis given to topics not commonly intended. Developmental complexity variation was obvious by different patterns in topic sequence and inclusion in the curriculum across the years of schooling. Cognitive complexity variation was revealed by relative emphasis on different performance expectations, that is, on what students were expected to be able to do.

Figures 2-1 and 2-2 have been chosen to illustrate the diversity in commonly intended topics and performance expectations across the six countries. They demonstrate the relative emphasis textbooks gave the common intentions found in curriculum guides. A topic treatment's clarity, focus, and coherence is seen not only in the emphasis given the topic but by its context among other topics. The relative emphases given various aspects of whole numbers in Figure 2-1 illustrate differences in topic complexity. Considerable emphasis given the meaning of whole numbers and the properties of their operations relative to whole number operations themselves would be considered by many to reflect greater complexity than emphasizing operations over the other aspects of whole numbers. In Figure 2-1 France and, to a lesser extent, Spain display this generally believed more complex pattern. Japan, Norway, Switzerland and the US placed relatively more emphasis on whole number operations as compared to the other two aspects thus suggesting a less demanding or complex pattern.

Figure 2-2 illustrates differences in cognitive complexity in Population One science textbooks. France and Japan clearly showed, across all topics, more complex expectations than the others. These two countries placed considerable emphasis on either 'understanding complex information' or 'identifying questions to investigate' as well as 'understanding simple information'. There was little, if any, emphasis in any countries' textbooks on 'designing investigations'.

WHAT DIFFERENCES EXIST IN MAJOR TOPIC CATEGORIES ACROSS THE COUNTRIES?

Beyond specifying particular topics, each national education system pursued broad topic areas reflecting their particular curricular vision and aims. Even at this level of broad content categories, variation among countries was obvious across topic categories and in the emphases within categories. Table 2-3 presents the percentages of Population One and Population Two mathematics textbooks allocated to seven broad mathematics content areas across the six countries. Table 2-4 presents similar data for six broad content areas in Population One and Population Two science textbooks. These categories come from the highest, most global levels of the TIMSS' curriculum frameworks. They sketch broadly the relative emphases on different topics as judged by relative emphasis in textbooks. As discussed earlier, textbooks represent the potentially implemented curriculum. Many textbook sections may never be covered. In other cases, topics are emphasized that are not reflected in the textbooks. However, content emphases in textbooks generally reflect the most likely content coverage possibilities sufficiently well to indicate the likely range of variation in topic emphases.

*Figure 2-1. Variation in the Distribution of Commonly Intended Topics
in Population 1 Mathematics Textbooks Across Six Countries*

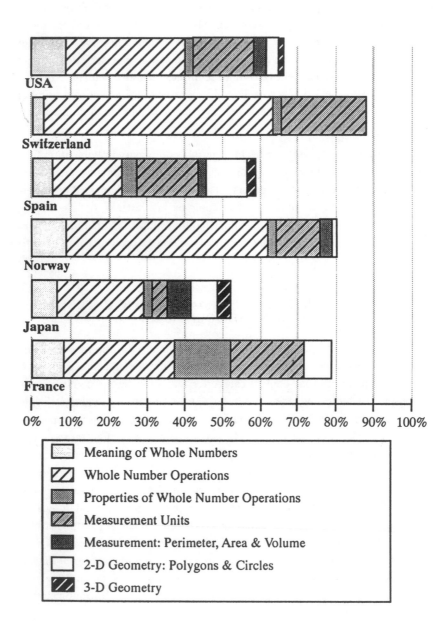

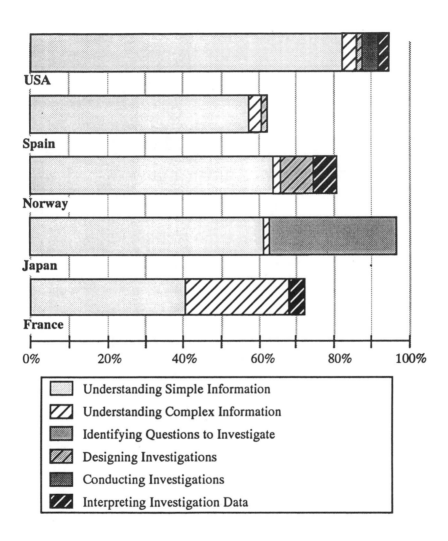

Figure 2-2. Variation in the Distribution of Commonly Intended Performance Expectations in Population 1 Science Textbooks Across Six Countries

Table 2-3. Percentage of Textbooks Devoted to Major Mathematics Framework Categories

Population 1

	Numbers	Measurement	Geometry	Proportionality	Functions, Relations, and Equations	Data Representation, Probability, and Statistics	Other Content
France	63	20	12	1	3	0	1
Japan	65	11	16	0	6	2	0
Norway	70	17	4	1	3	5	1
Spain	42	16	31	0	0	0	6
Switzerland	74	23	0	0	0	0	3
USA	62	18	9	0	2	9	12

Population 2

	Numbers	Measurement	Geometry	Proportionality	Functions, Relations, and Equations	Data Representation, Probability, and Statistics	Other Content
France	43	9	28	7	8	4	2
Japan	2	0	45	0	45	7	1
Norway	41	11	25	2	21	6	2
Spain	54	1	6	0	45	0	2
Switzerland	21	16	32	6	24	3	1
USA	35	10	16	6	28	13	10

Table 2-4. Percentage of Textbooks Devoted to Major Science Framework Categories

Population 1

	Earth Sciences	Life Sciences	Physical Sciences	Science, Technology & Mathematics	Environmental & Resource Issues	Other Science Topics
France	12	39	19	23	3	5
Japan	23	40	37	0	0	0
Norway	7	56	28	0	7	1
Spain	14	48	16	5	11	5
USA	14	38	26	9	5	8

Population 2

	Earth Sciences	Life Sciences	Physical Sciences	Science, Technology & Mathematics	Environmental & Resource Issues	Other Science Topics
France	36	9	48	3	4	1
Japan	23	25	52	0	0	0
Norway	7	63	17	2	7	4
Spain	16	28	41	7	1	8
Switzerland	11	57	19	4	8	1
USA	12	23	25	17	9	13

Clearly from Table 2-3 there were large differences in emphasis devoted to various broad mathematics topic areas among SMSO countries. Population One textbooks in all countries devoted substantial portions of their space to Numbers — although Spain's textbooks emphasized this topic less and other topics more than textbooks in the other countries. Textbooks in Spain and, to some extent, in France and Japan gave significant portions of their space to Geometry topics.

Textbooks at Population Two reveal even greater diversity in emphases than those from Population One. Japan and Spain placed greater emphasis on the typically more complex topics of Algebra (functions, relations, and equations). The other countries emphasized aspects of Number and Measurement which were virtually absent in Japanese textbooks. The particular aspects of topics — for example, Numbers — addressed in textbooks also reveals differences in topic complexity. More than 70% of French and Spanish textbooks treatments of Numbers include number subtopics such as 'number properties', 'number theory', and 'exponents'. Norway, Switzerland, and the US devoted 60% or less to these subtopics at Population Two. 'Number theory' was the only Numbers subtopic addressed in Japan's Population Two mathematics textbook. The implications of these differential emphases will become more evident in considering developmental complexity.

Similar diversity was demonstrated in the six countries' science textbooks. All Population One science textbooks devote more space to 'Life Science' topics than any others. The Japanese textbook also devoted considerable attention to 'Earth Sciences' and 'Physical Sciences'. The French textbook demonstrated a unique emphasis on 'Science, Technology, and Mathematics'.

French, Japanese and Spanish textbooks had comparatively more treatment of physical sciences content in their Population Two textbooks. Norway's and Switzerland's textbooks concentrated on life sciences topics at this level. The US textbooks at both population levels demonstrated a more unfocused treatment of topics devoting a little space to each major category.

WHAT IS UNIQUE IN THE INTENDED MATHEMATICS AND SCIENCE CURRICULA ACROSS THE COUNTRIES?

Each national education system also had unique emphases of both broad and specific topic areas reflecting their particular curricular vision and aims. Some of these patterns suggest greater complexity and demands on students than do others. These unique emphases may be examined through relative emphases in textbooks as was done for other aspects of curricular complexity. The following figures identify each country's five most emphasized textbook topics. Beside each topic is the relative percentage of textbook space devoted to treating that topic. Each figure also shows the total portion of textbook space devoted to the entire set of five most emphasized topics. Larger total portions suggest a more focused and in-depth treatment of a more limited set of content.

These figures highlight variation in topic complexity across the six countries. Neither the list of topics nor the emphasis given to common topics were identical across countries. Further, the five most emphasized topics accounted for significantly different portions of textbook space across countries. For example, 'animal types' was the most emphasized topic in Population One textbooks in Norway and the United States. However, Norway's textbooks devoted almost 17% of its content to this topic but the US not quite 7%.

The five most emphasized topics comprised as little as 20% of Population Two science textbook space in Spain and the US, but almost 70% of the textbook in Japan. The range in Population Two Mathematics textbooks was similarly less than 50% for Norway and the US but over 90% for Japan. Having the five most emphasized topics comprise a large portion of textbook space indicates a more detailed treatment of content. Covering fewer topics in greater detail in the same number of pages is an important aspect of topic complexity and depth.

The five most emphasized topics in mathematics textbooks from France, Japan, and Spain for both Populations One and Two demonstrated comparatively greater topic complexity. The four "not intended" (i.e., unshaded) topics in the French Population Two textbook gives further insight into the comparatively great emphasis given to Numbers at this level as indicated in Figure 2-3. Textbooks from France and Spain demonstrated similar comparative emphases on 'exponents, roots and radicals'.

In a similar manner as indicated in Figure 2-4, some of the five most emphasized topics in science textbooks from France, Japan, and Spain address what many consider more complex topics — 'chemical and physical properties of matter', 'descriptions of chemical changes', and 'atoms, molecules and ions'.

Figure 2-3. Percent Coverage of Most Emphasized Topics in Each Country's Mathematics Textbooks

Population 1 Mathematics

Rank	France	Japan	Norway	Spain	Switzerland	USA
1	Whole Number Operations — 31	Whole Number Operations — 24	Whole Number Operations — 54	Whole Number Operations — 18	Whole Number Operations — 63	Whole Number Operations — 28
2	Measurement Units — 20	Decimal Fractions — 18	Measurement Units — 13	Measurement Units — 14	Measurement Units — 23	Measurement Units — 12
3	Properties of Whole Number Operations — 14	Common Fractions — 10	Meaning of Whole Number — 8	2-D Geometry: Polygons & Circles — 13	Systematic Counting — 8	Other Content — 9
4	Decimal Fractions — 10	Measurement: Perimeter, Area & Volume — 7	Data Representation & Analysis — 4	Decimal Fractions — 10	Set Theory & Axiomatic Systems — 2	Common Fractions — 8
5	Meaning of Whole Number — 6	Patterns, Relations & Functions — 6	Patterns, Relations & Functions — 3	2-D Geometry: Basics — 9	Properties of Whole Number Operations — 2	Data Representation & Analysis — 6
Total	**81**	**65**	**82**	**64**	**98**	**63**

Population 2 Mathematics

Rank	France	Japan	Norway	Spain	Switzerland	USA
1	2-D Geometry: Polygons & Circles — 16	Equations & Formulas — 39	Equations & Formulas — 16	Equations & Formulas — 22	Equations & Formulas — 27	Equations & Formulas — 19
2	Relation of Common & Decimal Fractions — 11	Geometry: Congruence & Similarity — 23	2-D Geometry: Polygons & Circles — 11	Rational Numbers & Their Properties — 11	Measurement: Area, Perimeter & Volume — 22	Data Representation & Analysis — 7
3	Exponents, Roots & Radicals — 8	2-D Geometry: Polygons & Circles — 11	Measurement Units — 8	Common Fractions — 8	2-D Geometry: Polygons & Circles — 9	Other Content — 7
4	Properties of Whole Number Operations — 7	2-D Geometry: Basics — 10	Whole Number Operations — 7	Whole Number Operations — 7	Geometric Transformations — 8	Percentages — 5
5	Properties of Common & Decimal Fractions — 6	Data Representation & Analysis — 7	Common Fractions — 6	Exponents, Roots & Radicals — 5	Measurement Units — 7	Measurement: Area, Perimeter & Volume — 5
Total	**48**	**90**	**48**	**53**	**73**	**43**

Commonly Intended Topics in Shaded Cells

Figure 2-4. Percent Coverage of Most Emphasized Topics in Each Country's Science Textbooks

Population 1 Science

France		Japan		Norway		Spain		USA	
Reproduction of Organisms	12	Plant & Fungi Types	22	Animal Types	10	Rocks & Soils	17	Animal Types	7
Human Biology & Health	10	Physical Properties of Matter	16	Plant & Fungi Types	8	Light	7	Plant & Fungi Types	6
Influence of Science & Technology on Society	9	Animal Types	11	Life Cycles of Organisms	8	Human Nutrition	7	Organs & Tissues	5
Organs & Tissues	9	Bodies of Water	10	Electricity	6	Organs & Tissues	7	Earth in the Solar System	5
Science Applications in Math & Technology	8	Earth Physical Cycles	10	Sound & Vibration	5	Sound & Vibration	6	Science Applications in Math & Technology	4
	48		69		37		44		27

Population 2 Science

France		Japan		Norway		Spain		Switzerland		USA	
Earth Building & Breaking Processes	10	Electricity	10	Human Nutrition	19	Atoms, Molecules & Ions	17	Human Biology & Health	15	Science Applications in Math & Technology	7
Light	10	Chemical Properties of Matter	10	Human Diseases	17	Classification of Matter	13	Organs & Tissues	10	Rocks & Soils	3
Physical Properties of Matter	9	Weather & Climate	9	Human Biology & Health	15	Physical Properties of Matter	7	Life Cycles of Organisms	6	Organs & Tissues	3
Electricity	9	Descriptions of Chemical Changes	9	Organs & Tissues	13	Cells	6	Plant & Fungi Types	5	Animal Types	3
Rocks & Soils	8	Organism Energy Handling	8	Organism Energy Handling	6	Organs & Tissues	4	Human Diseases	4	Bodies of Water	3
	46		46		70		47		40		19

Commonly Intended Topics in Shaded Cells

WHAT VARIATION EXISTS IN TOPIC COMPLEXITY FOR THESE CURRICULA AS A WHOLE?

The preceding displays portrayed curricular topic complexity for two specific grade levels. Similar analyses here consider developmental complexity — the number of topics introduced, emphasized, and dropped from the curriculum each year. All countries require some mathematics and science instruction at nearly every grade level. However, as evident in Figures 2-5 and 2-6, the number of topics covered at any given grade varied among countries. Some countries focus on fewer topics at the same time, others on more.

Two distinct patterns emerge when comparing the data from Figures 2-5 and 2-6. Over all years, Japan and Spain intend to cover fewer topics each year and begin to remove topics earlier. Norway and the United States intend to cover a greater number of topics each year, to remove fewer topics and remove them later.[7] Japan's and Spain's approaches seem to allow greater focus and coherence in curriculum as it emerges across the years of schooling as well as within any single year.

Countries also differed in the sequence in which they present topics. The interaction of different education system structures, different content emphases, and different performance expectations for students within topics, created markedly varied patterns of topic coverage over the course of schooling. Variation in developmental complexity —the patterns of topic introduction and emphasis across the years — contributed to the overall portraits of national curricula. These portraits give impressions of the coherence and focus with which topics were treated by the curricula in the six countries. Figures 2-7 and 2-8 present comparative data for selected mathematics and science topics across the six countries. The topics in these displays were chosen to represent the diversity in topic coverage patterns across the six countries. They also provide an overall context for certain topics encountered in discussing topic and cognitive complexity and the classroom observations.

---

[7] In the next chapter, the suggestion is made that this may reflect a more child-centered, "students will get it when they get it" approach to teaching and learning. The practical result of this approach may, however, be redundancy and a lack of focus in instruction.

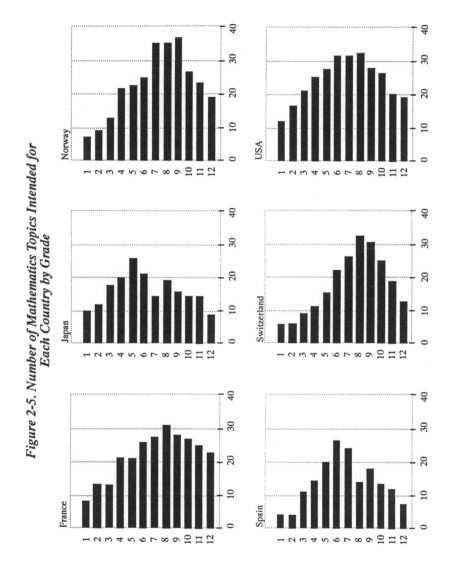

Figure 2-5. Number of Mathematics Topics Intended for Each Country by Grade

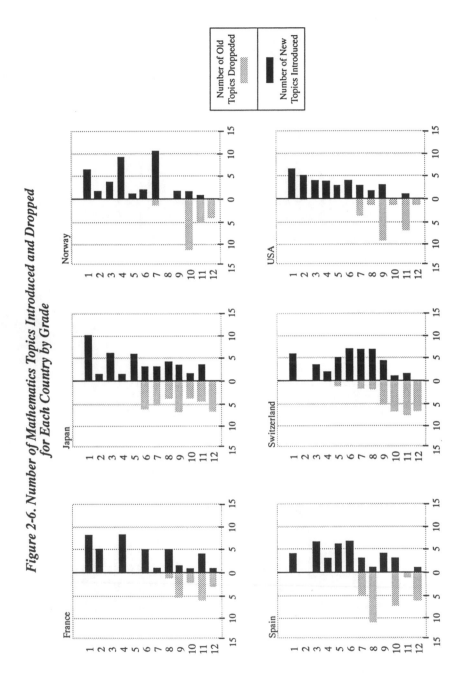

Figure 2-6. Number of Mathematics Topics Introduced and Dropped for Each Country by Grade

Figure 2-7 suggests at first a general impression of consensus across the six countries that some topics are more appropriately the focus for younger students and others more appropriate for older students. However, within this general consensus, variation is readily apparent. Japan and the US focused on the 'properties of whole number operations' earlier and completed coverage of this topic before either France or Norway. Japan and Spain focused on the 'relation of common and decimal fractions' earlier than the others. Japan intended completing this topic before Switzerland even began it (Moser, 1992a). Japan's and Spain's pattern across the first three examples was one of comparatively short focus. Example 4 exemplifies several of developmental complexity's main variations — sustained emphasis and focus over many years (Japan), sustained consideration with intermittent or no true focal points (US and Norway), comparatively moderate consideration and emphasis (France), and relatively brief consideration with no emphasis (Spain). The way these different developmental patterns are woven together across topics to form a curriculum have profound implications for students' learning.

Figure 2-8 shows less consensus about the appropriate age for topic consideration. The general impression is one rather of a lack of consensus. Example 1 demonstrates a pattern in which France, Norway, Spain, and the US began considering topics as many as eight years earlier than Japan, or Switzerland. Across three of the four topics found in Figure 2-8, Switzerland demonstrates a consistent pattern of relatively late introduction but, consequently, rather focused emphasis. Considering the years of topic coverage together with the points of topic emphasis reveals two distinct patterns — sustained, unemphasized treatment and a relatively more emphasized treatment. Norway and the US demonstrated the first pattern, one consistent with an integrated approach to science (demonstrated also in other ways in their curricula)[8]. The relatively more emphasized topic treatment approach was demonstrated by the other countries and was consistent with the practice of most to offer separate courses in different sciences as early as the Population Two level.

---

[8] In Norway, this picture is largely a reflection of the organization of their curriculum guides. Norway's curriculum frameworks are written in three year cycles. Teachers have the option of when over this three year period they will teach an intended topic. This is facilitated by the fact that teachers often stay with the same group of students for as many as two of these three year cycles (see the Norway case study in Part II of this volume). This organization and use of the Norwegian guides practically precludes emphasis of a topic at any particular grade level.

*Figure2-7. Curriculum Coverage for Selected Mathematics Topics
Across Student Ages*

Example 1: Properties of Whole Number Operations

Country	\<Student Age\>												
	6	7	8	9	10	11	12	13	14	15	16	17	18
France	-	-	-	-	-	-	-	-	-	-	-		
Japan		•	•	•	-								
Norway			-	-	-	-	-	-	-	-			
Spain			-	-	-	-	-						
Switzerland		-	•	•	•	-	-	•	•				
USA	-	-	-	•	•								

Example 2: Relation of Common & Decimal Fractions

Country	\<Student Age\>												
	6	7	8	9	10	11	12	13	14	15	16	17	18
France				-	-	-	-	•					
Japan			-	-	•	•							
Norway					-	•	-	-	-	-			
Spain					•	•	-	•	-				
Switzerland							•	-	-	-			
USA			-	-	-	•	•	-					

Example 3: Exponents, Roots & Radicals

Country	\<Student Age\>												
	6	7	8	9	10	11	12	13	14	15	16	17	18
France						-	-	•	-	-	-	•	
Japan							•		•	•	-		
Norway								-	-	-	•	-	-
Spain						-	-	-	•				
Switzerland								-	•	•	-	-	
USA						-	-	•	•	-	-	•	

Example 4: Equations & Formulas

Country	\<Student Age\>												
	6	7	8	9	10	11	12	13	14	15	16	17	18
France								-	-	-	•	-	
Japan			•	•	•	•	•	•	•	•	•	•	
Norway	-	-	-	-	-	-	-	-	-	-	-	-	-
Spain						-	-	-	-	-	-		
Switzerland	-	-	-	-	-	-		•	•	-	•	•	
USA	-	-	-	-	-	•	•	•	-	•	-		

Note: Ages 9 and 13 are TIMSS Student Populations 1 and 2

- topic covered in curriculum   • topic emphasized in curriculum

Figure2-8. Curriculum Coverage for Selected Science Topics Across Student Ages

Example 1: Earth Building and Breaking Process

	Student Age												
Country	**6**	**7**	**8**	**9**	**10**	**11**	**12**	**13**	**14**	**15**	**16**	**17**	**18**
France			–	–	–	–	–	•	–	–	•	–	
Japan								–	•	•	–	–	
Norway					–	–	–	–	–	–	–		
Spain						–	–	–	•	–	•	•	
Switzerland								–	–	•	•	–	
USA	–	–	–	–	–	–	–	–	•				

Example 2: Organs and Tissues

	Student Age												
Country	**6**	**7**	**8**	**9**	**10**	**11**	**12**	**13**	**14**	**15**	**16**	**17**	**18**
France				–	–	•	•	•	•	•	•	•	
Japan			–	–	–	–	•	•	–	–	•	•	
Norway					–	–	–	–	–	–	–	–	
Spain			–	–	•	•	–	–	–	–	•	•	
Switzerland								–	–	•	•	–	
USA	–	–	–	–	–	–	–	–	•				

Example 3: Reproduction of Organisms

	Student Age												
Country	**6**	**7**	**8**	**9**	**10**	**11**	**12**	**13**	**14**	**15**	**16**	**17**	**18**
France		–	–	–	–	•	•	•	–	–	•	•	
Japan					•				•	•			
Norway		–	–	–	–	–	–	•	–	•	–		
Spain				–	•	–	–	–	•	–	•	–	
Switzerland			–	–	–	–	–	–	•	•	–	–	
USA	–	–	–	–	–	–	–	–	•				

Example 4: Chemical Properties of Matter

	Student Age												
Country	**6**	**7**	**8**	**9**	**10**	**11**	**12**	**13**	**14**	**15**	**16**	**17**	**18**
France								–	–	•	•	•	
Japan					–	•	–	•	•	–	•	•	
Norway					–	–	–	–	–	–	–	–	–
Spain						–	–	•	–	•	–	•	
Switzerland							–	–	–	•	•		
USA						–	–	–	–	–	•		

Note: Ages 9 and 13 are TIMSS Student Populations 1 and 2

- *topic covered in curriculum* • *topic emphasized in curriculum*

WHAT VARIATION EXISTS IN COGNITIVE COMPLEXITY OF THE MATHEMATICS AND SCIENCE CURRICULA ACROSS THESE COUNTRIES?

Displays and discussion so far have mainly addressed topic and developmental complexity. Similar variation can be seen in cognitive complexity — in what is expected of students — as was demonstrated in Figure 2-2. The following two figures present textbook percentages devoted to each of the broadest performance expectation levels from the TIMSS' frameworks. In each case, the five categories represent major types of performances that might be expected of students learning mathematics or science.

Each broad category represents a different type of requirement for students learning and understanding. For example, in mathematics a student could simply memorize multiplication facts to satisfy a 'knowing' expectation. He or she might learn to repeat without thinking a procedure for multiplying two two-digit numbers together (a 'routine procedure' expectation). However, different, more-difficult-to-attain understanding of multiplication would be needed to demonstrate why that procedure gave a correct answer when multiplying two two-digit numbers (one aspect of 'mathematical reasoning'). Similarly, 'understanding definitions' of scientific vocabulary seems less demanding than 'theorizing, analyzing and solving problems' by recognizing applicable scientific principles or by constructing, interpreting and applying a model. The following displays show significant variation across the six countries in what students are expected to do as part of learning mathematics and science. Variation in the cognitive and performance demands placed on students calls for substantively different learning environments and experiences. To the extent that these differences are present in national curriculum and pedagogical implementations, qualitative differences in student learning among countries should be expected.

Figure 2-9 and 2-10 illustrate the diversity among broad performance expectation categories. There is similar diversity within each category as well resulting in the significantly different potential learning experiences for students in the six countries. Paralleling the greater topic complexity (discussed earlier) in French, and Spanish textbooks, was a more complex pattern of expectations in the Population Two mathematics textbooks from these countries. 'Mathematical reasoning' and 'investigating and problem solving' received comparatively greater emphasis in these countries' textbooks. Further, less than 40% of space in French and Spanish textbooks was devoted to the less demanding 'knowing' and 'using procedures' expectations compared to at least 60% for the other countries.

*Figure 2-9. Population 2 Mathematics Textbooks' Relative
Emphases of Performance Expectations*

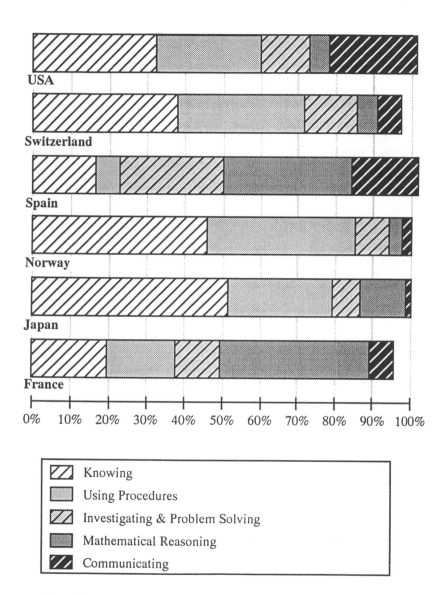

NOTE: *Percentages across all categories may exceed 100% due to
possible mutiple codings.*

Figure 2-10. Population 2 Science Textbooks' Relative Emphases of Performance Expectations

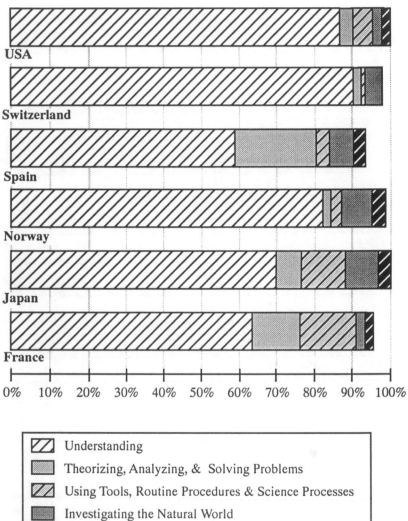

Comparable differences were seen in Population Two science textbook performance expectations. Textbooks from Norway, Switzerland, and the US devoted more than 80% of their space to 'understanding'. In contrast, 70% or less was devoted to this single expectation in textbooks from France, Japan, and Spain. French and Spanish textbooks exhibited a comparatively stronger emphasis on 'theorizing, analyzing, and solving problems' while French and Japanese textbooks also exhibited a considerable emphasis on 'science processes'. The patterns of performance expectations in textbooks from France, Japan, and Spain seem to demand that comparatively more complex performances be mastered.

Figures 2-11 and 2-12 illustrate a similar contrast within the science 'understanding' performance expectation at both Populations One and Two. The graphs indicate a strongly dominant emphasis on 'understanding simple information' at Population One. The French textbooks stand alone in having emphasized 'understanding complex and thematic information' almost as much as 'simple information'.

At Population Two Japan stands out as having emphasized 'understanding complex and thematic information' more than 'simple information'. French and Spanish textbooks also exhibited considerably more emphasis on 'understanding complex and thematic information' than those from Norway, Switzerland, and the US, with comparatively less emphasis on 'understanding simple information'. The French, Japanese, and Spanish textbooks seem to have emphasized mastering more demanding performances in science learning than did the textbooks from the other countries.

Figure2-11. Population 1 Science Textbooks' Relative Emphases on Understanding Simple and Complex Information

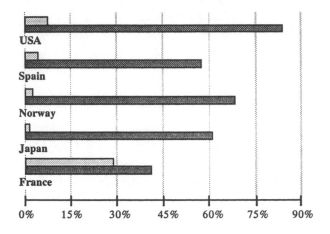

Figure2-12. Population 2 Science Textbooks' Relative Emphases on Understanding Simple and Complex Information

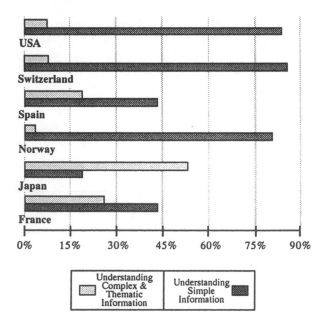

**Figure2-13. Population 1 Mathematics Textbooks' Emphases on
Aspects of Whole Numbers**

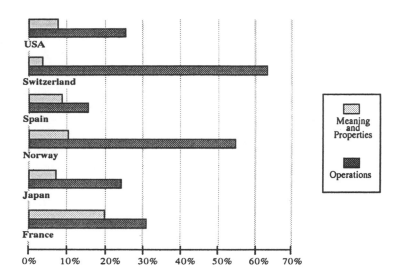

**Figure2-14. Population 2 Mathematics Textbooks' Emphases on
Aspects of Common & Decimal Fractions**

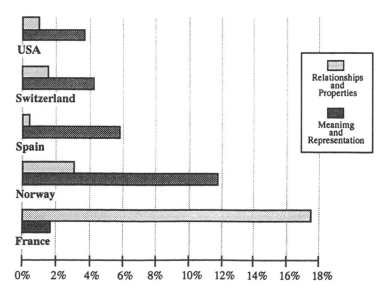

Topic and developmental complexity may also have an impact on the intended cognitive complexity encountered by students. As noted earlier, some topics are more basic but straightforward than others. A curriculum rich in these more basic topics would likely be less complex and demanding compared to one emphasizing less basic topics. This can be seen by comparing mathematics curricula for Populations One and Two across the six countries. Much of each country's Population One curriculum was devoted to 'whole numbers'. However, as Figure 2-13 reveals, there were considerable differences in emphasis given different aspects of this topic. Textbooks from France and Spain exhibited almost as much emphasis on the 'meaning and properties of whole numbers' as on 'whole number operations'. This seems more demanding — both in topic complexity and, perhaps, cognitive complexity.

Similarly, in Population Two, there were considerable differences in emphasis given different aspects of 'common and decimal fractions' as Figure 2-14 shows. Japan is conspicuously absent from this display. Comparing it to Figure 2-7 shows why – Japan completed all intended coverage of these topics two years earlier. In fact according to these data, Japan completed all intended coverage of this topic the year before Switzerland began to address this topic. In Population Two, France and Spain emphasized this topic. It was also the culmination of consideration of this topic in France's curriculum. The pattern of completing this topic at Population Two or earlier, together with the relative topic complexity demonstrated in Figure 2-14, suggests that the French and Japanese curricula offered more complex and demanding opportunities in this content area than the others.

This variation in emphasis just discussed is particularly noteworthy when combined with a consideration of what expectations textbooks emphasized for these contents. The commonalities in topic complexity in mathematics curricula, evident in the list of intended topics from mathematics curriculum frameworks and textbooks presented in Table 2-1, appear less similar after considering the great variation in how these topics were treated. Pedagogical strategies and approaches varied even in presenting the same or similar content, as the next set of figures makes clear. Focusing on any one topic, that topic's array of broad performance expectations varies considerably among countries.

Considering topic complexity and cognitive complexity together portrays more fully the character of typical learning experiences students encounter. Figure 2-1 and 2-2 earlier exhibited variation in emphasis on commonly intended topics and performance expectations in textbooks across the six countries. Taken together these displays suggest even more strongly that students' learning experiences even for the same topics potentially should be qualitatively different among the six countries profiled. Not only were there emphasis differences for the same topics, there were also emphasis differences for the same expectations for students' learning. Comparing the two figures suggests that one country may not only address more complex topics but expect more complex performances to be mastered by their students for these topics. The comparative complexity among national curricula could be compounded.

The following figures begin to bring to light this compounded complexity by portraying the relative emphases textbooks have on student performance expectations for particular topics. Both topics in Figures 2-15 and 2-16 were emphasized in the respective textbooks (see Table 2-1, page 36). Considerable differences in what Population One students were expected to do with 'whole number operations' is evident from Figures 2-15. This is a basic mathematics topic which receives considerable textbook emphasis at this level (see Figures 2-1, page 40). Figures 2-1 showed considerable emphasis on the 'meaning of whole numbers' and the 'properties of whole number operations' in some countries compared to emphasis on 'whole number operations'. This was true in textbooks from France, and Spain.

Figures 2-15 compounds this previous portrait. Even for 'whole number operations' textbooks from France and Spain showed more emphasis on 'mathematical reasoning'. Japan, Norway, and Switzerland almost exclusively emphasized the two performances 'knowing and using vocabulary' and 'using routine procedures'. The textbooks from Spain and the US exhibited more varied expectations.

Figures 2-16 shows similar emphases in Population Two textbooks on 'equations and formulas'. Consistent with the similarities in Population One textbooks, French and Spanish textbooks exhibited a comparatively strong emphasis on 'mathematical reasoning', and those from Japan and Norway almost exclusive emphasis on 'knowing and using vocabulary' and 'using routine procedures'. Spanish and US textbooks again showed comparatively more varied performance expectation emphases. The Swiss textbook also demonstrated more varied expectations for this topic at this level in contrast to the Population One expectations for 'whole number operations'.

Figure 2-15. Distribution of Performance Expectations in Population 1
Textbooks for Whole Number Operations

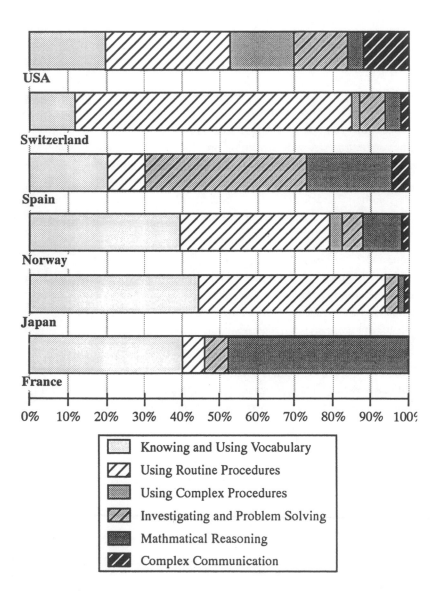

*Figure 2-16. Distribution of Performance Expectations in Population 2
Textbooks for Equations and Formulas*

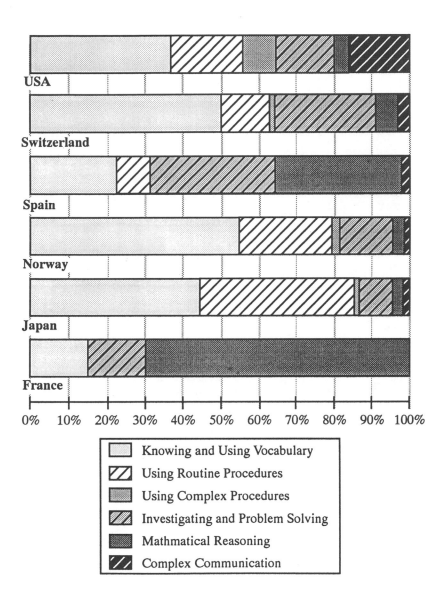

Figure 2-17. Distribution of Performance Expectations in Population 1
Textbooks for Organs & Tissues

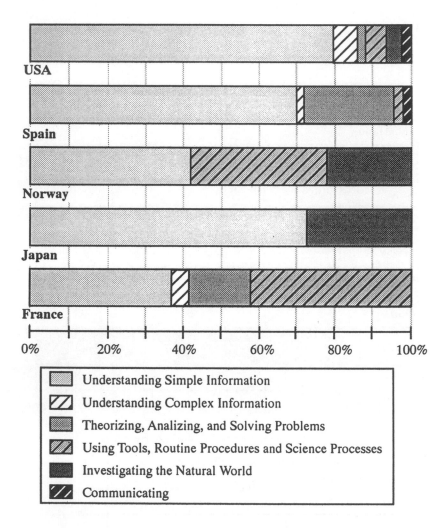

Figures 2-17 and 2-18 draw a picture for science similar to that for mathematics in Figures 2-15 and 2-16. Both science topics in Figures 2-17 and 2-18 were commonly intended at their respective student population levels (see Table 2-1, page 36) and exhibited considerable differences in what students were expected to do. 'Organs and tissues' was one of the five most emphasized topics for textbooks in France, Spain and the US (see Figures 2-4, page 47). However, textbooks from France and Spain uniquely contained the expectation 'theorizing, analyzing and problem solving' practically absent in the other countries' textbooks. Similarly, textbooks from Japan and Norway uniquely contained the expectation 'investigating the natural world'. These textbooks treated 'organs and tissues' quite differently. Although in almost all countries the predominate expectation for this topic is 'understanding simple information', the French and Spanish textbooks additionaly include an academic or conceptual approach while the texts from Japan and Norway additionaly include a more process-oriented, experiential approach. The US pattern suggests a focus primarily on facts and information with no other major approach present.

The portrait from Population Two science textbooks is somewhat different. 'Chemical properties of matter' was one of the five most emphasized textbook topics only in Japan (see Figures 2-4, page 47). Japanese textbooks are distinguished by their predominate emphasis on 'understanding complex information'. Analogous to the expectations in their Population One textbooks for the topic 'organs and tissues', the French and Spanish textbooks uniquely contained the 'theorizing, analyzing, and problem solving' expectation which was practically absent from other countries' textbooks for this topic. The Norwegian expectation profile was essentially identical to that displayed in their Population One treatment of 'organs and tissues' with an almost equal emphasis on 'understanding simple information', 'using tools, routine procedures and science processes' and 'investigating the natural world'. Again, this suggests an approach to science rich in experiential exploration. US and Swiss textbooks were similar in their almost exclusive emphasis on 'understanding simple information', respectively devoting more than 80% and 60% to this single performance expectation. This suggests a primary focus on facts and information, although this was not as pronounced in Switzerland as in the US.

Figure 2-18. Distribution of Performance Expectations in Population 2 Textbooks for Chemical Properties of Matter

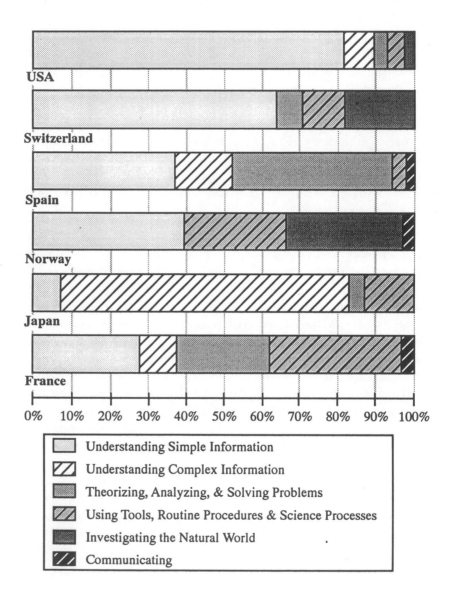

SUMMARY

This chapter focused on curriculum frameworks and textbooks. The portraits of mathematics and science topics developed here may be considered background for the main story about classroom instruction which is developed in the next chapter. As far as these curricular materials served as classroom resources and helped shape learning activities, their differences suggest that students' learning experiences were considerably different across these six countries. There was both commonality and diversity in topic, developmental, and cognitive complexity of school mathematics and science among these countries. However, given the diversity presented in this chapter, these portraits portend the likelihood of significant differences in what is observed in classrooms from country to country.

Distinct portraits of these six countries' mathematics and science curricula emerge from this chapter's data. The curricula in France, Spain and, to some extent, Japan seem to have been more involved and posed more varied demands on students than did those in Norway, the US, and, to a lesser extent, Switzerland. In Japan this came from relatively complex topics and treating them in a comparatively focused manner. In France and Spain this was achieved by combining topic complexity and cognitive complexity — both by selecting complex or theoretical aspects of topics and expecting students to deal with this substance in more demanding ways. The more diffuse, sustained treatment of topics in Norway and the US suggests curricula that were less focused and, perhaps, typically demanding less complex performances from students. The Swiss curricula appeared more focused than those of Norway or the US, but Swiss textbooks appeared at times quite similar to Norwegian textbooks and at other times quite similar to US texts.

What do these differences imply for teaching and learning? What impacts might one expect to see in these countries' classrooms? What types of similarities and differences among the countries should be expected inside their classrooms? Perhaps the differences here are simply surface differences which do not reflect major underlying differences in thinking about and approaching school mathematics and science. Alternatively, perhaps these curricular differences actually reflect significantly different approaches to understanding and teaching mathematics and the sciences. The significance of the differences discussed in this chapter will become more clear through Chapter Three's discussion of classroom observations.

Chapter 3:
The Classroom Story Unfolds: Observing the Implementation of Curricular and Pedagogical Intentions

Students enter their eighth grade mathematics class in the United States stopping to check their paper "leaves" in the "garden" displayed on the back wall of the classroom to see how much of the "leaves'" area has been destroyed by "insect activity" since the previous day. While students are recording their "leaf" data, the teacher puts assignments on the board: leaf project, expression simplification, and graphing equations. These projects were begun several days ago and each is to be completed on different days the following week.

As class begins, the teacher asks students how many have not completed the expression simplification worksheet. None have finished it so the teacher tells them to form groups of two or three to complete the worksheet. Students arrange themselves in groups and begin to discuss how to simplify and then solve each algebraic expression.

As the teacher moves about the room monitoring students' progress, she realizes that some students do not understand how to complete the exercise. She asks all students who are having difficulty to join her in a group and she goes over the procedures with them. Once she has explained several exercises to the group, she lets them continue without her and resumes moving from group to group monitoring the students' activity. After a brief time, she announces to the class that they will end this activity in a few minutes and move on to something else. She tells students that she will be in the room during lunch next week if they need further help in completing this assignment.

Later, with about 15 minutes remaining in the period, the teacher has students return to their seats and to turn their attention to the screen at the front of the room. The teacher enters an equation into her graphing calculator which then is projected onto the screen. The teacher leads the class in predicting how the graph of various equations will look. She asks students to tell something about the geometry of the lines the equations represent, how two lines are related to one another, and the difference in orientation of two lines represented by two different equations.

Students appear to be thinking and often offer voluntary responses to the teacher's prompts. This discussion continues until the signal is heard ending class.

In France, after correcting homework on the blackboard with the help of several students, the teacher began her eighth grade mathematics class by announcing the lesson's title: "expansion and factorization". She explains that expansion basically requires applying the distributive rule and after illustrating this with an example has a student demonstrate with another example. After this she explains that factoring is the opposite operation and illustrates this with an example. Having explained one example, the teacher has students work on two further examples.

After a brief time, the teacher has a student solve the two exercises on the blackboard. The teacher makes a few comments and then moves on to more complicated examples that entail using both expansion and factorization. She demonstrates how to work with this type of exercise and then has students work on a similar example. The teacher moves about the classroom monitoring students' progress and indicating their errors.

After three minutes, the teacher asks a student to put the exercise on the board. The student writes the exercise on the board clearly labeled with the three steps presented earlier by the teacher: expand, factor, and proof. The teacher discusses the two exercises students have worked on and then assigns a few exercises to be completed for the next lesson.

Classroom activities are a dynamic interaction between subject matter content, teachers, and students. Teachers use broader curriculum goals and intentions to select or create pedagogical activities and exercises packaged into lessons for students. This chapter explores similarities and differences found in mathematics and science classroom lessons across the six SMSO countries. Subject matter differences as well as age and grade differences figure prominently in this story. Both will be seen to contribute to important qualitative differences in lesson format and delivery within and across the six countries.

CHARACTERISTIC PEDAGOGICAL FLOW

The heart of the story appears simple, almost self-evident. Classroom practices really do differ considerably among countries. In observing classrooms from each SMSO country, the practices observed often did not conform to the expectations or assumptions held by experts from other countries.

Events common in one country often seemed strange and unusual to persons from another. A teacher in Norway sent students into the woods behind the school to gather specimens for a lesson on classification while the teacher remained in the room. A teacher in Japan used a videotape of botanical specimens made earlier on an class field trip as the basis for a classroom classification lesson. Those not from France were intrigued by French teachers' practice of giving students explicit instructions about which notes and diagrams to include in their notebooks to ensure that all students recorded fundamentally the same information. A class size of only 14 students in a Swiss elementary mathematics classroom seemed quite small to some representatives, while 35 students in a Japanese elementary mathematics classroom seemed rather large. Those not from the United States found the fairly common practice of students exchanging and grading each others' homework surprising, and many on the research team were intrigued by one homework assignment from Spain in which students constructed a water pump at home.

As interesting and significant as these differing practices may be, they are not sufficiently fundamental to account for the qualitatively different nature of the lessons observed. Something deeper and more profound is involved. The SMSO team believes that interesting differences of the sort described above are not random variations in a widely shared, consensual view of how to construct lessons. Rather, these differences reflect particular stances toward pedagogy and classroom activity that vary among countries and, at times, even within a country. These stances are expressed in certain recurrent patterns of instruction and classroom activity. The SMSO team coined the term "characteristic pedagogical flow" (CPF) to label these recurrent patterns in a set of lessons.

"Pedagogical" is part of the term because at issue is the pedagogical strategies and approaches typical of a set of lessons. "Characteristic" is included because the SMSO team firmly believes that certain pedagogical strategies are enacted repeatedly in a country's classrooms because they are characteristic of a wide-spread perspective on students, teaching, learning, subject matter content, strategies developed during teacher preparation and training and in the shared professional lives of teachers — in short, all the factors that interact to determine the typical in a given classroom context. "Flow" is included in the term to indicate that applying characteristic pedagogy to an instructional situation is at best only partly analytical. Most teachers do not consciously work through a planning process or routines that order and shape classroom learning activities. Rather, teachers "flow" through a familiar activity — teaching in a certain setting — based on their past experiences, training and beliefs. The "rules" governing their activities, the values guiding their choices, the repertoire of activities and routines from which they choose are largely below the conscious level for most teachers — especially during the course of the lesson. All planning and execution of plans have some conscious decision making, but for extremely familiar activities — those for which the constraints and possibilities are well known — decision making can be embedded seamlessly into carrying out that familiar activity with little awareness of its component phases and choices.

Others have labeled this sort of experience as a "flow" experience (Csikszentimihalyi, 1990; Schiefele & Csikszentmihalyi, 1994)[1]. Here the SMSO team has appropriated the term because it seems to describe so well teachers' experiences while teaching. Retrospectively or when called on to do so, most teachers can think analytically about what they usually do in classrooms. At the time and spontaneously, experienced teachers seem rarely to do so. CPF, characteristic pedagogical flow, was chosen as a term combining these three essential elements — a focus on enacted pedagogical approach, a search for the typical and characteristic, and a recognition that actual classroom interactions are rarely analytic at the time for experienced teachers.

"Characteristic pedagogical flow", and the stance or perspective that underlies it, is derived from complex — but often unexamined and subconscious — interactions of particular conceptions of subject matter and pedagogy (i.e., methods and strategies based on a view of the teaching/learning process). "Characteristic pedagogical flow" involves, among other factors, the interplay of three key attributes — content representation and complexity, content presentation, and the nature of the classroom discourse accompanying content presentation. Highlighting these three attributes foreshadows one important lesson that emerged from analyzing the lessons observed: content is key. Having begun the study by focusing on content delivery in each of the countries involved, this finding may not seem surprising. However, the centrality

---

[1] Csikszentmihalyi describes "flow" as "optimal experience". He states that descriptions of "flow" experiences indicate that they include "a sense that one's skills are adequate to cope with the challenges at hand, in a goal-directed, rule-bound action system that provides clear clues as to how well one is performing. Concentration is so intense that there is no attention left over to think about anything irrelevant, or to worry about problems. Self-consciousness disappears, and the sense of time becomes distorted" (1990, p. 71). Teaching may well not be a flow experience for many teachers. However, for experienced, informed and prepared teachers in settings with adequate freedom and resources, the conditions for a flow experience seem often to be met. Further, intense concentration and a lost sense of time characterize much of teaching. When teaching becomes such an "optimal experience", it seems to involve little on-the-spot analytic thinking but proceeds more as a result of planning and intuiting, as a pure expression of training, experience, beliefs, perceptions and preparation.

of content waxed and waned in the team's thinking, finally to emerge as deeply, integrally, and surprisingly involved in coming to understand pedagogy cross-nationally.

It now appears that differences in how subject matter content is construed are the basis for most of the differences in instructional practice evident in these cross-national comparisons reported here. The degree of coherence evident in a lesson, for example, varied among countries and reflected, to a large extent, differences in how teachers thought about content. Those who viewed content as an unfolding narrative attended to its "story line" — its sequence and increments — in a much more systematic way than those who viewed it as discrete encounters with important topics. Similarly, how teachers structured discourse around content appeared to vary as a function of how they represented content to themselves and to their students.

Stodolsky, in **The Subject Matters** (1988) reached a similar conclusion regarding the teaching of mathematics and social studies in the US. She found that views of subject matter heavily influenced the way elementary teachers organized lessons and selected activities. The SMSO project's findings verify and extend that conclusion in searching for "characteristic pedagogic flow" cross-nationally, with differing curricula, and at differing ages.

This chapter's qualitative analyses elaborate on this notion that subject matter matters by showing how the same subject matter can be conceptualized, presented and discussed in dramatically different ways in different national settings. One point made in Chapter Two was that the seemingly plausible ideas that "mathematics", "school mathematics", or "mathematics education" were culturally invariant are not validated as one investigates more deeply the substance of what educators in specific locales think ought to be taught a particular grade level's students.

A growing research literature within the US and other countries disputes the notion that subject matter teaching at the elementary or secondary level is more alike than different across disciplines and ages. If this is true within countries, it is hardly surprising to find it true across countries. There were examples of mathematics and science instruction in all six countries that seemed oddly familiar to those accustomed to particular approaches to instruction. France, Norway, Spain, and the US all evidenced elements of traditional, teacher-centered, "transmission" instruction. France, Japan, Switzerland, Spain, and the US demonstrated elements of Socratic dialogue in some lessons. Some observations from Japan, Norway, Switzerland, and the US suggested a child-centered philosophy and approach shaped at least some aspects of lessons and lesson activity.

Lessons shared some similarities across countries and observations within any one country often demonstrated variation. Even so, representatives from the six countries involved detected consistent patterns in the data that indicated essential — not accidental or superficial — differences in how mathematics and science lessons were organized in the six countries. Thus, not only did subject matter matter, at least for the countries considered here, but the nature or character of lessons appeared to differ more among countries than within countries. That is, lessons within a country tend to exhibit a consistent CPF.

The remainder of this chapter elaborates the CPF concept by (1) presenting and illustrating three key dimensions of CPF, (2) highlighting some of the key CPF issues found in mathematics and science lessons at the two student age-levels considered, and, finally, (3) concluding with a description of CPF for each of the countries, a characterization based on the observed lessons and the key CPF issues they raised.

CPF: THE INTERACTION OF CONTENT AND PEDAGOGY

A lesson's pedagogical flow is shaped by the inextricable manner in which curriculum and pedagogy interact to yield learning experiences of a particular quality. Lessons differ because of qualitatively different ways in which teachers' conceptions of lesson content interact with their use of instructional activities. Pedagogical commonalities in lessons indicate similarities in teachers' concepts of content and use of instructional activities which are, in short, manifestations of CPF. In the model of educational experiences introduced in Chapter One, CPF is an interaction between two aspects of the implemented curriculum, content and instructional practices. These are represented by the two boxes at the classroom level labeled "Teachers' Content Goals" and "Instructional Practices". Figure 3-1 shows how these key aspects of CPF are related to Chapter One's model.

The content aspect of CPF is addressed in the model with the question, "What are students expected to learn?" This issue was explored in the previous chapter by discussing various aspects of content complexity — topic, developmental, and cognitive complexity — as embedded in curricular guides and textbooks. The same issues may also be explored from classroom observations. The classroom observations, however, provide a better opportunity to explore issues of content representation, presentation and discourse. These aspects of CPF are addressed in the model with the question, "How is instruction organized?" The heart of CPF, the interaction between these two aspects, is represented in the model by the double-ended arrow. The CPF arrow spans the column associated with the question, "Who delivers the instruction?" This makes it clear that the heart of CPF — the interaction of content and instructional activities — is a function of teachers, their practices, their background, training, and their ideas about students, subject matter, learning, and teaching.

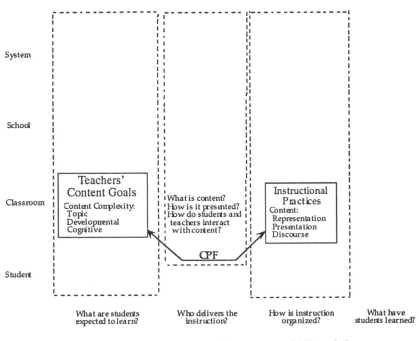

Figure 3-1. The Relationship Between CPF and the Educational Experiences Model

It is important to distinguish between the product of CPF, an observable characteristic of lessons, and the proposed process behind CPF, a function of the ideas, beliefs, theories, and pedagogical repertoires held by teachers. The discussion of CPF here is primarily and deliberately descriptive. It attempts to make clear what it is about lessons that differs fundamentally and gives rise to characterizations of qualitative differences. Little is said about the specific ideas or theories teachers may have which inform their choices and actions and which are manifested in their classroom lessons. The SMSO project focused on instructional practices because of its charge to aid in TIMSS instrument development. It did not investigate in depth teachers' ideas or theories or seriously attempt to solve the difficulty of linking particular ideas or theories to specific practices. There remains, however, an inseparable link between the described results of CPF and the proposed underlying process, even when current data are insufficient to illuminate that link.

The approach here is that lessons' characteristic quality may be explored in part by asking three questions: What is the lesson's content? How is the lesson content presented? How do students and teachers interact with each other and the lesson content? Investigating these questions will help clarify what it is about the interaction of content and instructional practices that give lessons their characteristic qualities.

CONTENT COMPLEXITY AND REPRESENTATION: WHAT IS THE LESSON'S CONTENT?

Content complexity and content representation are two aspects of a lesson's content contributing to its character. The former relates primarily to curriculum and was discussed extensively in the previous chapter. The latter is primarily related to instruction. However, before considering either further, how observable a lesson's content is is something that must be considered. Content in the observed lessons ranged from being almost entirely hidden — virtually a complete lack of content-laden interactions — to being markedly obvious as content was thoroughly and explicitly embedded in classroom interactions and activities.

In some lessons, content was essentially invisible — little could be said about lesson content because so little content was obvious from the observational reports. "Invisible" content lessons were those in which students worked individually on projects, worksheets, or exercises — typically with little overt interaction between students and the teacher. Invisible content was more often seen in mathematics lessons than science lessons although it was seen in both types. In these lessons the teacher usually did not address the entire class for instructional purposes but might begin the lesson with administrative comments or a review of relevant procedural directions. These lessons appeared to focus primarily on specific routines or procedures students were to carry out rather than on any particular subject matter or topic. Interactions between teacher and students consisted of questions or comments about procedural processes rather than substantive issues.

As overt interactions focusing on subject matter increase, classroom observations become relevant for issues of lesson content complexity and representation. Content complexity (discussed in the previous chapter) relates to perennial issues in psychology and education. Conceptual (i.e., topic) complexity and developmental complexity help characterize the lesson's content, while cognitive complexity helps characterize teachers' and students' interactions with content. Cognitive complexity relates to language about the lesson's content during classroom activity.

Topic complexity suggests simply that some topics are more difficult than others. Some of these differences in difficulty were discussed in Chapter Two but it was also evident in the observations. For example, an elementary science lesson observed in Norway concerning what happens in nature in the Spring seems less difficult than a lesson observed in a US classroom regarding the nature of electricity in terms of atoms' structure and functions. Obviously this comparison does not imply that the observed Norwegian lessons were systematically less difficult than those observed in the US. The example merely serves to illustrate the relative difficulty of two particular lessons.

Developmental complexity suggests that the same topic can be addressed with increasing difficulty, elaboration and refinement in a sequence of lessons — particularly a sequence spanning several years. This idea underlies Bruner's (1966) "spiral curriculum" – students repeatedly return to the same topic but encounter it in increasingly rich, complex and more inclusive forms.

Developing the idea of classification in science lessons is an example of increasing developmental complexity. From a first two-fold classification of things as "living" and "non-living", later lessons may proceed to classify living things as "plants" and "animals", and continue with the development of ever more elaborate and complex taxonomies. This may proceed to the borders of classifiable things and to sophisticated taxonomic issues, for example, how to classify viruses and other things that begin to blur the original distinction between living and non-living things.

It seems obvious that content's developmental complexity should be related to, if not identified with, children's cognitive developmental levels. However, this is not necessarily true. Content development in increasingly complex forms requires examining many related lessons in several grades or years. The SMSO observations are cross-sectional, focusing on two grade levels. This precludes direct characterization of developmental complexity and any comment would require extensive inferences that a country's children are exposed to similar experiences in a certain grade, even when children of different ages reach that grade in different years, with different teachers, and in changing educational settings. A few observations, however, led to some comments about the perceived appropriateness of lesson subject matter for students in the observed grade — a reflection more of the related but distinct issue of children's cognitive development rather than the developmental complexity of the curriculum. These observations will be discussed later.

An important aspect of lesson content more directly observable has to do with how content is represented in classroom explanations, illustrations, and discussions. Most topics can be considered in ways that more or less closely relate to students' immediate, concrete experiences. Close or loose ties to concrete experiences can be developed by many different pedagogical activities. Explanations may use readily understood analogies to things in students' everyday experience, or a more abstract approach to explanations may be used. Illustration of a principle, concept or phenomena may draw on students' everyday experience, or a more abstract approach based on the interconnections of the subject matter may be used instead. Ultimately all good instruction relates lesson content to some aspect of students' experience, to things they already know. Those connections in school lessons may come from life experiences, specially generated concrete classroom experiences, or previous classroom experience with other aspects of a topic in comparatively more abstract form. Good lessons make connections; the kinds of connections vary. This variation in kinds, methods, and timing in discussing connections help give lessons very different character. Systematic observation of similar approaches across sets of lessons at a grade level in a country help reveal characteristic pedagogy and are aspects of the characteristic pedagogical flow in that nation's classrooms.

Briefly comparing two geometry lessons for 13-year-olds provides an illustration. In one Swiss lesson, the teacher had students form a circle and then, with string, form the sides and diagonals of a series of polygons noting for each the number of sides, corners, and diagonals. Students then worked in small groups to find a general rule that described the relationship among these numbers.

In a French lesson, the teacher had students evaluate the sizes of parts in several different geometric figures. The students needed to draw on their knowledge of definitions and rules for how the sizes of various parts of figures related. As students discussed the figures, it became apparent that the teacher had assigned impossible values to at least one part of each figure. This became obvious because the assigned values violated the rules describing the relationship among the parts of the figures.

The Swiss lesson created a common concrete experience on which to base further discussion. The French lesson called on previous classroom experiences with formal definitions and rules for specific mathematical objects. Both lessons connected to student experiences but did so in very different ways.

CONTENT PRESENTATION:
HOW IS THE LESSON'S CONTENT PRESENTED?

Content presentation, or, more precisely, the strategy of presenting content during the course of a lesson, is another dimension of CPF to consider. This involves how content is encountered by students during a lesson and how content is distributed within a lesson by teachers' choices of pedagogical strategies. This certainly relates to lesson content, the first dimension considered. The strategy for presenting a topic in classroom activities is considerably affected by how complex the topic is, how fully it is to be represented at a particular point in a curricular sequence, and the teacher's conception of the topic and of the appropriate learning experiences for that topic. Certainly, however, the topic content which students are to encounter and the classroom activity strategy by which the topic will unfold for them in a particular lesson are distinct.

The pedagogy question, "How is the content presented in a lesson?", examines the logic and connection of a lesson's development, what activity segments will be used in the lesson, and how these segments fit together in a reasoned whole — an aspect of a lesson referred to elsewhere as its coherence (Prawat, 1989). Some lessons exhibited steady, focused consideration of a single topic through one or more related activities. Other lessons involved a collection of ideas and methods. Aspects of content in the former lesson type seemed to cohere more closely compared to content elements in lessons of the latter type. Two elementary mathematics lessons illustrate this contrast.

In one Japanese lesson observed, the teacher presented the class with two sets of numbers, one an exact count of the previous day's attendance at soccer games and the other an estimate of the previous day's attendance at baseball games. The entire lesson dealt with discovering which method of rounding (i.e., by tens, hundreds, thousands, etc.) could be used with the soccer numbers to show that the previous day's soccer game attendance was greater than the baseball attendance. Different activities took place during the course of the lesson but all related to this single, central problem.

In another lesson, from the United States, students were given oral addition practice, practice with other arithmetic operations at the blackboard, a review of telling time along with a related worksheet, and an exploration of metric measurements by reading and answering questions from the student textbook. This lesson dealt with varied mathematical content. It also made use of varied types of activities. There was no clear, central content for the lesson and thus there could be no integrated set of activities to unfold this central content. The choice of which activity was used for which content in the lesson seemed arbitrary and if any pedagogical principle underlay the entire lesson it would be on the order of "students should experience a variety of activities within a lesson", perhaps with the intention of holding their attention longer.

Further, the way a particular activity is implemented can yield qualitatively different learning experiences. Consider the case of seatwork — that is, of students working individually at their seats during a lesson period on some tasks assigned by the teacher. Such seatwork was used very differently among countries' observed lessons. In Norway and the US, seatwork was often a single block of independent student practice usually comprising a major portion of a lesson. In observations from France, Japan and Spain, seatwork was distributed throughout the lesson as several brief periods of focused independent student work in a context of topic development led by the teacher. This development drew immediately on the personal experiences, successes, and difficulties of individual students during each seatwork segment of the lesson's sequence.

CONTENT DISCOURSE ISSUES:
HOW DO STUDENTS AND TEACHERS INTERACT WITH EACH OTHER AND THE LESSON'S CONTENT?

The final CPF dimension considered concerns how students and teachers interact with each other and with a lesson's content. This includes the kinds and levels of language used and how content is present in this language. It also includes what kinds of discourse — which interactions and discussions — are typical and encouraged. This is obviously and intimately related to the previous CPF dimensions. Similar student and teacher activities in different lessons can involve content in significantly different ways. Homework — work done by students outside of class as part of a teacher's planned experiences for those students — can be a part of in-class interactions very differently in different lessons, even when a broad characterization of those lessons would indicate that all dealt with homework. For example, homework can be reviewed solely to check whether students' work is right or wrong or it can serve as an opportunity to review, recapitulate, and summarize the important ideas from a previous lesson. In the former case, content was only tangentially important and the focus was rather on assessing student performance and, more indirectly, maintaining student engagement in studying the subject in general (since rarely is such student grading of homework used for one aspect of content and not another — rather it is a part of general classroom routine and teacher expectations). In the latter case, homework was used to bring students' interactions with content into the classroom context for further subject-oriented discussion. Even if such use of homework was also a matter of classroom routine and teacher expectations, it was a structure designed to focus teacher and students on subject matter. Homework did not have as its purpose evaluating students to assign marks. Homework review in the former lesson engages neither teacher nor student with content other than superficially. The second situation directly seeks to foster engagement with content by both teacher and students, and to make outside work a part of continued, in-class content development. As this example shows, the same lesson activity can be implemented differently because of varying pedagogical perspectives and thereby differentially engage students. (Farnham-Diggory, 1994, develops a similar argument.)

This relates to cognitive complexity as discussed earlier. This concept comes from cognitive psychology and has been a central concern for those emphasizing "teaching for understanding" — for integrated, empowering student assimilation of subject matter. "Teaching for understanding" stands in contrast, for example, to rote learning of a "transmitted" body of knowledge that is assimilated and connected only to a skeletal representation of that content knowledge in students' minds, with any connections beyond the body of knowledge coming later. Several have attempted to define different types of cognitive demands within a universal hierarchy, e.g., Bloom's taxonomy (Bloom, Engelhart, Furst, Hill, & Krathwohl, 1956). The point here, however, is that different cognitive demands require different responses from students, engage different student resources, lend different qualities to learning experiences and, most likely, result in different types of learning. Further, teachers' perceptions of the cognitive complexity of various tasks have marked effects on their characteristic pedagogy and on the tasks used to shape and support fruitful classroom interactions.

The cognitive demands placed on a student — that is, the thinking a student is required to do, can be quite different depending on the type of question asked by a teacher or the nature of an assignment's required tasks. Examples from several mathematics classrooms for 9-year-olds illustrate some of these cognitive demand differences.

In one US lesson on measurement, the teacher asked the class a series of questions: how many millimeters in a centimeter? how many centimeters to a meter? and so on. Similarly, several Norwegian lessons began with the teacher orally presenting to the class a series of exercises dealing with basic operations for which they were to perform mental calculation or recall learned number facts: six times four...; eight times four...; two times three...; and so on. Students in both of these examples were expected to retrieve from memory correct responses to the teacher's questions — little, if any, thought other than simple memory retrieval was required or expected.

In one lesson observed in Spain, the cognitive demands placed on students were quite different. The teacher began by drawing two pictures on the board, one containing 10 flowers and the other containing six flowers. He then asked the students, "What is the difference between these two bunches of flowers?" and had them draw both a copy of the two original flower bunches and a third representing the difference between the two bunches. In this example the teacher, dealing with basic arithmetic operations similar to those in the previous examples, required students to produce a visual representation of a subtraction problem and to draw a new picture representing its solution. Certainly students could simply retrieve the solution from memory. However, the teacher's presentation of the problem reminded students that they could reason their way to a solution and use meaningful representations and associations in doing so. The approach continued to reinforce the value of concrete representations even when they were not essential to the underlying task. The effect of this approach may not be immediate for the lesson's content but, considered cumulatively, may create different approaches, attitudes, and independence in students.

This type of approach was the substance of an entire lesson observed in Japan. The teacher began a lesson on measurement by presenting a picture depicting a tape measure $1^1/_5$ meters long. The teacher asked the class, "How do you convert the length of a tape which measures $1^1/_5$ meters into centimeters? Try to explain your method by using either an equation or a diagram or whatever you can." Beginning in this manner and continuing through the entire lesson, the teacher asked students to represent their solution in some manner other than a simple numerical answer, and to explain not only their result (solution) but the reasoning they used in arriving at it. The focus was on "how" to answer such questions, rather than on "what" particular answers were. Representations and approaches — and students' spontaneous choices of them — played a more essential role in this lesson. These examples illustrate how even very similar questions in lessons can be approached differently by teachers and how this may contribute to qualitative differences in student learning experiences.

THE CPF CONSTRUCT: A SUMMARY

CPF is a hypothesized construct that integrates the pedagogical beliefs and approaches of teachers which underlie their actual classroom conduct. It integrates pedagogical and curricular choices which, it is hoped, also reveals approaches that are characteristic of sets of lessons observed in particular countries, at particular grade levels and for particular broad contents. CPF embeds the effects of the decisions teachers make about which topics to present, how to present them, and how to organize and implement lessons. Pedagogical planning and, in particular, classroom conduct may be so automatic and routine for teachers that they often are "flow" experiences in which decision making and action are reflexive rather than reflective. Even so, CPF embodies the results of those typical and characteristic decisions and reflexive actions. It encompasses the complexity and representation of a lesson's content, how the content is presented to and encountered by students, and how the teacher and students interact around the lesson's content. These three dimensions are conceptually and empirically important in creating the portraits of characteristic teaching from the observed lessons presented here.

The next sections of this chapter present discussions of CPF issues specific to each of the two student age groups and each subject matter, i.e., Population One Mathematics, Population Two Mathematics, Population One Science, and Population Two Science. Following this is a brief discussion of CPF issues for each SMSO country. These country-level characterizations, for the most part, attempted to integrate the four categories of lessons previously discussed (Population One Science, etc.). The chapter concludes with a few summary remarks on CPF after this extended synthesis of empirical work.

CPF ISSUES IN POPULATION ONE MATHEMATICS

Common topics addressed by all SMSO countries' observations for Population One mathematics included addition and subtraction, measurement and measurement units, place value, and rounding. How visible topics were in lessons differed because classroom activities and teacher-student interactions differed.

In some classrooms specific subject-matter content was almost invisible. Lesson topics were identified only by examining the textbook or worksheet with which students worked. In these lessons, students worked mostly alone and independently to accomplish assigned tasks from the text or worksheets. Interactions between students and teachers were primarily organizational or managerial —which problems to do, when tasks were to be completed, etc.

Lesson content was more visible when students asked for clarification or assistance in completing assigned tasks. These were often brief exchanges on proper solution procedures for the assigned problems or tasks. Content in these observations was still usually presented by textbook or worksheet. The teacher may have briefly introduced and explained the task on which students subsequently worked alone (or occasionally with a partner).

Some lessons were comprised of a single cycle of these two segments — introduction/explanation and individual work — and other lessons by repeated cycles of these two types of segments. In either case, students spent most of the lesson working independently. This lesson type was observed mostly in Norway and the United States. Such lessons usually addressed only one or two closely related topics but in one US lesson students were directed to three different tasks in quick succession and the tasks involved more than four different mathematics topics.

Other lessons had a structure similar to the above but content was more visible because of the nature of the tasks and the interactions among students and teacher. The amount of interaction between students and teachers was not necessarily more than lessons of the earlier type, but interactions seemed to be more conceptually- and content-oriented.

For example, in one Spanish lesson the teacher presented students with a series of practice calculations by having them imagine themselves engaged in various farm-related transactions — buying or selling products, determining prices, determining the amount of money and change required, etc. In a Swiss observation, students worked in pairs using sets of wooden cubes to represent various division and multiplication situations. The majority of Swiss and Spanish observations were of these types. Although students worked largely alone or in pairs in these lessons, their interactions with teachers focused on curricular content more often than did lessons featuring completing assigned text exercises or worksheets.

Other lessons were composed almost entirely of interactive activities. Content became fairly apparent through observing these interactions. In these lessons one activity or a sequence of several closely related activities developed the lesson's content. For example, in one US lesson about a coordinate system, students worked in pairs to construct diagrams on paper, one student giving movement directions which the other recorded. In a Japanese lesson students were encouraged to develop and explain several different ways of determining how much of a tape measure measuring $1^1/_5$ meters is left if $3/_5$ meters are used up. Only a few observed US lessons were of this type while virtually all of the observed Japanese lessons were.

Traditional "drill and practice" was evident in observations from Norway, Spain, Switzerland, and the US. In the Swiss and Spanish observations, a single 'big' principle usually served to organize lesson activities and interactions although multiple problems were used to explore and develop this single focus. Both multiple topics and multiple exercises were typical in Norwegian and US observations. There were no French observations at this age level.

Use and emphasis on a 'weekly plan' was widely observed in both Norwegian and Swiss classrooms. In both countries, teachers created this weekly plan, put it on the board, and periodically reminded students of the tasks involved. However, students were expected to demonstrate initiative and take responsibility for their own learning in carrying out this plan. There were no instances of teachers lecturing students in the lessons from Norway and Switzerland. The weekly plan and the accompanying emphasis on student responsibility and the expectation that students would more independently pursue their own learning seems to partly explain this absence of lectures.

Lessons in Japan and the US often were also organized around a single 'big' principle or idea. However, the interaction between students and teacher differed in these lessons. Such interaction was almost constant throughout Japanese lessons and constituted most of the content development. Teachers guided content development through carefully controlled conversation about a single situation or problem. Students were explicitly directed to discover multiple methods for solving or thinking about the situation or problem being considered. The conversational style was understood and familiar. This allowed the content of the conversation to focus entirely on the problem at hand. In similar US lessons, lesson content was an activity the teacher had carefully planned, but not necessarily of a type which students consistently encountered. Teacher and students interactions often dealt more with activity-related issues than with the conceptual content to be developed by the activity.

CPF ISSUES IN POPULATION ONE SCIENCE

Specific content was the CPF dimension that contributed most to qualitative differences in the character of lessons in Population One science classrooms. In Population One mathematics classrooms, lessons typically addressed topics from a core common to all SMSO countries. Even when this was not so, there was a recognition that topics addressed by only some of the countries would be addressed at some point in the mathematics curriculum of the other countries. By contrast, the list of science topics addressed in Population One science lessons differed greatly among countries. That is, there were differences in what constituted science across countries. This was particularly evident at this level in spite of the extensive commonly intended topics found in Table 2-1.

For example, some Norwegian and Swiss science lessons contained content which in the US would be considered social science (e.g., geography), along with the natural science content categorized in the TIMSS science framework. Science in the European tradition embraces both the natural sciences and elements of the social sciences such as history, anthropology and geography. This is reflected in the science content of classroom lessons. In Switzerland, for example, no subject called 'science' is taught at this level. Instead, what is taught is "Natur – Mensch –Mitwelt" (Nature – Human Being – The world we live in) — that is, human environmental studies. This content is built around children's everyday experiences and involves exploring objects and phenomena common from children's local environments. Lessons integrate aspects of many different disciplines in discussing and exploring topics such as food production and preparation, housing, gardening, fuel for warmth, and how people in earlier times experienced these same things. Topics are adapted to children's experiences and ways of thinking, rather than shaped by the formal structure of disciplines from the sciences. Topics are presented using descriptive, common language rather than scientific, academic terms. This distinct approach to Population One science was instrumental in Switzerland's decision not to participate in the TIMSS Population One science testing.

These different conceptions of science contribute to a qualitatively different character across the lessons observed. Content in US and Japanese observations were restricted to natural science topics. Observations from Switzerland and Norway revealed more social science influence. No observations were made in Population One science classrooms in either France or Spain.

Several countries' SMSO representatives commented on the comparatively brief time allocated to TIMSS science topics at this level (e.g., in Norway, Switzerland, and the US). This relative lack of emphasis on science may not only reflect curricular differences but also the reticence of teachers who often did not feel adequately prepared or equipped to deal with science topics. There were, nonetheless, explicit goals for students' learning of science at this level in each of the SMSO countries (see Chapter Two).

Norwegian and Swiss Population One science lessons appeared to have social goals broader than goals related to particular science topics. This likely results from the broader conception of 'science' noted above. In any case, these lessons addressed science topics while integrating issues of culture, language, and politics.

Several examples from Norway's lessons illustrate this. In one observed lesson, the teacher introduced a thematic unit in which students would prepare reports on various topics concerning nature, the environment, and conservation. The teacher introduced the unit by asking, "If you had the power and authority, what would you do to make it better for every living creature here where we live?" In two other lessons, topics were introduced by listening to a song and reading a poem.

Broader goals of identifying and collecting information from various sources and of constructing and communicating with others the results of one's research were evident in some observed lessons in Norway and Switzerland. Furthermore, the Swiss representative noted the existence of an unofficial "rule" in Switzerland that all science lessons at this level should also be exercises in language and speech development.

In the observations from Japan, Norway, and Switzerland, both the selection of science topics for lessons and the way these topics were developed were closely related to the everyday experiences and interests of nine-year-olds. Students' observations and conjectures were integral and important to these lessons. Whether the observed lesson was a hands-on activity or a historical exploration, student-generated explanations and hypotheses were emphasized. In contrast, US observations showed lessons more oriented to facts and definitions and for which the teacher served as the primary source of information.

CPF ISSUES IN POPULATION TWO MATHEMATICS

Strong similarities were observed in content and activities in Population Two mathematics classrooms across the six countries. Writing and simplifying equations, exploring exponents and powers, and basic geometry were topics commonly addressed in almost every countries' lessons. In virtually every lesson students spent some time working on exercises and were either assigned homework, reviewed homework or both. However, in the context of these similarities, even more remarkable differences were noted in how these elements were incorporated into lessons.

Homework was involved in observed lessons in one of three ways. In some lessons reviewing and correcting students' homework functioned as review of important concepts from a previous lesson and then led to further topic development. In these lessons homework and in-class homework review were integral parts of planned topic development. A topic was introduced, explained, and then students were given opportunity to apply what they had learned in one or two problems. Students' in-class work then served as the basis for further topic development. In this approach homework seemed an extension of in-class seatwork in which students practiced applying new concepts. This lesson approach was consistently observed in lessons from France and Spain.

A second, similar way homework functioned in a lesson was much the same as the first but only at times — that is, its use was optional or conditional. Homework was not reviewed consistently as an integral part of planned topic development, but appeared to be conditional on teacher perceptions of students' need for such review. This type of lesson was found in Switzerland and Norway.

In a few lessons, homework review centered on whether students had completed it and whether they had arrived at correct solutions for homework problems. Homework discussion and interaction did not concern conceptual or procedural issues from the homework. When homework completeness and correctness was determined, the lesson proceeded. Homework appeared at best to function as an indicator of readiness to move to another topic. These lessons were observed in the US and Norway.

One of the more striking differences among observed lessons was in presentation and development of the lesson topic. Two lessons clearly illustrate this contrast. In the first, the teacher began by asking students to estimate how thick a piece of paper would be if it was folded in half ten times. Students were then given pieces of paper to explore this. Because in practice it is difficult to fold a piece of paper more than five or six times, students had to derive some way of representing the relationship between the paper's thickness and the number of folds.

Students proposed various solutions to the teacher. Eventually, the teacher worked with the entire class and constructed a table which had columns for the numbers of folds, layers of paper, and measurements of the folded paper's thickness. Together teacher and students investigated the relationship between the number of folds and the number of layers for 10 folds and 30 folds. They finally generalized that the number of layers in n folds could be written as 2^n.

In the other lesson, the teacher had asked students to prepare word problems based on their recently acquired knowledge of powers. One student presented his problem by writing on the blackboard: "Given: We have a giant piece of paper whose thickness is 0.1 mm. How many times must one fold this piece of paper so that it reaches the height of the Eiffel Tower, which measures approximately 300 m?"

The teacher asked students how this problem could be solved and worked with the class to find a solution. After writing an expression for the paper's thickness after one (0.1×2 mm), two (0.1×2 mm $\times 2 = 0.1 \times 2^2$ mm), and three folds (0.1×2^3 mm), the relationship between the number of folds and the paper's thickness was quickly generalized to n folds: 0.1×2^n mm. Therefore, the expression to be clarified was $0.1 \times 2^n > 300{,}000$ (converting the Eiffel Tower's height from meters to millimeters).

A student recalled the general rule that if $a \neq 0$, $\dfrac{a^{n}}{a^{m}} = a^{n-m}$

Applying this general rule, the following expressions were generated:

$$2^n > \frac{300,000}{0.1} \quad , \quad 2^n > \frac{3 \times 10^5}{10^{-1}} \quad \text{and} \quad 2^n > 3 \times 10^6$$

From experimenting with the calculator, the student knew that $2^{21} = 2,097,132$ and that $2^{22} = 4,194,264$. Because the teacher did not want to consider "partial folds", the student concluded that the solution was 22. The teacher then wrote on the board, "After 22 folds, the thickness of the folded paper is more than the height of the Eiffel Tower".

The Swiss lesson above proceeded from presenting a situation closely connected to students' experience and using their experience to develop a mathematical relationship inductively. The French lesson, in contrast, used formal mathematical definitions and theorems that were brought to bear upon a situation closely connected to students' experiences. Lessons from Switzerland and Japan were primarily of the former type while lessons from France were consistently of the latter.

In other lessons, content presentation appears to have been essentially determined by the textbook. There was some variation in the amount and distribution of practice exercises students were expected to do (i.e., seatwork) and in whether students worked individually or in pairs to complete the exercises. The amount of explanation provided by the teacher also varied. However, the distinguishing feature of these lessons was presenting mathematics from the textbook.

For example, during the topic's explanation or discussion in some lessons the teacher referred to the text, at times reading it. In one observation the teacher had students follow along in the text or consult it for correct answers, and in another the teacher had students read the textbook themselves. In the latter lesson, as students worked on exercises in the text, if they had questions about the topic they were often told to read specific areas of the text. In another observed lesson, when reviewing homework or work done in class, the teacher used the text explicitly as the source for the correct answers. In all these cases, the mathematics presented in the class appears as an authoritative body of knowledge to which students and teachers must hold. Lessons of this type were seen from Norway, Spain, and the US.

Reference to the 'weekly plan' was again seen in both Norwegian and Swiss lessons although not as extensively as in Population One lessons. Nonetheless, the use of a weekly plan along with the amount of time students spent working independently, the conditional use of homework, and the importance of students' experiences in developing mathematical concepts (particularly in the Swiss lessons) suggests that there continued to be an emphasis upon students' responsibility and independence in learning (EDK, 1995; Moser, 1992b).

CPF ISSUES IN POPULATION TWO SCIENCE

As for Population One science lessons, CPF issues for Population Two science also related to content. Here, however, rather than focusing on Population One's issue of what constitutes 'science', content complexity was a major source of qualitative differences among observed Population Two lessons.

Similar content topics were encountered in each of the SMSO countries — air pressure, magnetism, rocks and minerals, oxidation, the water cycle, electricity, and so on. The same content was not portrayed with equal complexity at this age level among the six countries. The same level of content might be reached in later grades but, in this cross-sectional investigation of one level, there were clear complexity differences. Observed French and Spanish lessons presented many of the same topics as the other countries but in substantially more complex forms.

These lessons also differed in how science was represented. Some lessons emphasized relatively more the processes of doing science — that is, making observations, describing what happens, making conjectures and hypotheses, and discussing and evaluating observations, descriptions and hypotheses. Other lessons gave relatively more emphasis to formal scientific definitions and principles applied to various phenomena.

In lessons of the first type, teachers often asked questions like "What do you see?", "What happens?", "What changes?" as well as "How do you explain what you see?" In lessons of the second type, teachers asked similar questions but tended to emphasize definitions of key terms and concepts, to identify appropriate principles or laws, and to apply or calculate specific scientific formulas for a specific situation or problem. Lessons of the first type were observed predominately in Switzerland and the US. Lessons of the second type were observed primarily in France, Japan and Spain.

The distinction here is descriptive. One lesson type is not considered to have a richer view of science than the other, but merely a different relative emphasis. Almost all observed lessons contained some of both kinds of content. Nonetheless, these relative differences in emphases did give qualitatively different character to the two types of lessons. Comparatively more emphasis on one of these approaches resulted in a very different experience of science. The comparative differences in emphases might be linked to students developing very different ideas about the nature of science and attitudes towards it. Students' understanding of scientific concepts might become quite different and contribute to differences in students' performance on items assessing science achievement.

CPF ISSUES IN THE SIX COUNTRIES

FRANCE

The most important aspects of CPF in France that emerged from these observations concerned teachers' relation to subject matter and the complexity of the subject matter presented. French teachers were comparatively highly involved with subject matter in their interactions with students. Content both in mathematics and the sciences was relatively more complex formally than for other SMSO countries. French mathematics and science instruction focused strongly on content and emphasized theoretical justification for mathematics procedures and scientific rules.

A lesson's focus material was often presented comparatively abstractly. The framework and main aims for many lessons were particular symbolic operations (e.g., expansion and factorization in mathematics) or scientific principles (e.g., the law of series in measuring voltage between close points). Teaching typically tended to follow a "demonstration-rule-example," content-oriented approach. That is, teaching formal principles was followed by ample opportunities to apply the knowledge just learned. This general approach was consistent with content representation moving regularly between the more abstract and the more concrete or applied.

Consistent with this, teaching was strongly directed by the teacher and, in the observed mathematics and science lesson, interactions assumed a heavy reliance on teacher expertise. The teacher typically directed the flow of information for students during the lesson, often explicitly directing them to record particular things in their notebooks or otherwise highlighting elements of the content presented.

The frequency with which teachers departed from a strict reliance on the textbook was evidence of classrooms centered on teacher expertise. In those instances, teachers either provided their own content examples or embellishments or made selective use of the student text material. The teachers' handling of content and the kinds of activities used left little doubt that teachers were the subject matter discipline experts in the classroom.

There was heavy reliance on precise language in the observed French lessons. Teachers spent significant amounts of time ensuring that students mastered definitions of specific terms during instruction. The teachers' frequent admonitions for students to record specific aspects of the lesson in their notebooks likely related precisely to this aspect of content coverage. It is wrong to assume that in these content-oriented, teacher-directed classrooms, interaction was solely between teachers and student. The observations also showed considerable student-to-student interaction, but only in the application or non-expository part of each lesson.

The French lessons followed a fairly consistent pattern regardless of subject matter (i.e., mathematics or science). The lesson structure had several segments. Lessons usually began with review of previous assignments, followed by a careful overview of the general content. This overview served as a kind of "advance organizer" for what followed. In some science lessons, particularly those introducing new material, this brief overview was by a teacher demonstration illustrating the new topic. After the overview, the material was examined in more detail. Lessons then typically proceeded to an expository part followed by an application part. In the expository part of the lesson, the goal was to arrive at formal disciplinary concepts (e.g., theorems and laws) and to develop relationships described by these principles — for example, how the effective voltage of a generator was related to the number of turns in the generator's coil.

Often in this exposition, students were asked pointed questions about the content presented. The teacher-student interchange was a very traditional (in many countries) pedagogical approach of "elicitation-response-feedback" — teacher questions elicited student response to which the teacher provided a response ("feedback") as part of the continuing topic development. This question-answer format is a common feature (e.g., in the US and other countries) of lessons by those who consider themselves fairly traditional, content-oriented teachers. In France, as in the US, this question-answer strategy has two purposes. First, it seeks to motivate students to attend to what is being said. Second, the specific question asked alerts discerning students to important parts of the content.

Lesson exposition in the French classrooms was not simply verbal. Blackboards frequently were used to present symbolic information — the blackboard was an integral part of the teacher's lesson presentation, particularly in mathematics. In science, various visual representations — diagrams, charts, drawings, demonstrations — were used to provide examples and make concrete what otherwise might be fairly abstract concepts. Some of these were an integral part of the curriculum material in the textbook.

After the lesson's expository phase, students were typically given opportunities to apply the information presented. The application phase, as was the exposition, was very structured and close-ended — intended to reach definite points by lesson's end. Students were clearly expected in these exercises to acquire a certain fluency in carrying out specific well-defined operations (e.g., computing voltage from various generator terminals, identifying rock strata given certain criteria, etc.).

The application phase mirrored and extended the exposition. Students were given the opportunity to carry out for themselves certain applications demonstrated during the exposition. Students often worked in pairs or small groups during this part of the lesson.

The typical lesson's final phase was a combined "debriefing" — that is, discussion and providing a teacher response to the exercise phase just completed — and an explicit foreshadowing of the next lesson's content. Daily homework assignments were central to this foreshadowing aspect of instruction. The amount of time devoted to homework review at the start of a lesson was a clue to the seriousness of homework's role in this continued sequence of developing new content.

JAPAN

Japan's national curriculum governs the content of lessons. The government approves all textbooks used and, in some cases, writes curriculum materials. Textbooks are recognized as an important tool for both teachers and students in transmitting information. Teachers are provided with detailed instructions on how lessons are to be taught.

The most important CPF feature in the observed Japanese lessons was the high level of involvement by both teacher and students with lesson content. Similar to French teacher's active direction of lessons, Japanese teachers actively sought to elicit explicit, fairly complex content from students. This content was not particularly formal. The Japanese teachers sought to foster understanding by causing students to consider multiple approaches in discussions and activities. Teachers frequently used discussion to bring out student responses that developed lesson concepts. In the process, teachers subtly focused these discussions on content principles.

"Hans", small groups of students commonly formed for discussion and activity purposes during lessons, were often used to further work on a problem after which each "han" was asked to present its results to the whole class. A "han" is a fairly stable student group changing only once or twice during the year and frequently employed during lesson discussions. After reports by each han, the whole class discussed the various solutions provided and eventually ended with the best solution for their purposes.

In this way, teachers actively encouraged students to discover and consider multiple approaches to situations and problems, both in mathematics and science lessons. However, as class discussion proceeded, students were carefully (but subtly) directed by the teacher towards the lesson's focal concept.

The Japanese lessons also used a fairly consistent lesson structure — what might be labeled problem-oriented teaching. Lessons began with the presentation of a problem or situation. Discussion was then used to help students generate ideas and approaches to the problem or situation. This discussion might include the student-to-student discussions in the hans. This was followed by general discussion of ideas and solutions subtly directed by the teacher. The general discussion ended with a succinct summary by the teacher at the lesson's end.

Science lessons were mostly activity-based, with the activity's objectives clearly stated by the teacher. Science lessons were very structured — teachers directed an initial discussion and group activities followed. A problem was stated by the teacher. Ideas from the students led to the concept which was the basis for further discussion or experimentation. A single lesson typically presented one concept, and allowed time for student discussion and reflection.

For example, in one Population One science lesson, the teacher began with a discussion on where rain water goes. Students raised their hands before offering ideas. The teacher recorded the ideas on the board. The teacher continued probing until many ideas about evaporation had been listed. Students were asked to write these ideas in their notebooks. They were asked to discuss with their group how they might design an experiment to test their ideas about rain water evaporation.

Adequate time was given for this phase of a lesson (not just in the specific observation above). This indicates the importance of process and not only content in Japanese science teaching. When groups have developed ideas for experiments, they were typically shared with the entire class. Classmates helped refine various groups' proposed suggestions. Teachers intervened in these discussions — for example asking questions concerning the need to control variables and how to do so. This general discussion, with subtle teacher contributions, shows how students refined their planning process.

This method helps students to understand that there are multiple ways to test the same hypothesis. As the preceding example and discussion suggests, science processes were heavily emphasized in teaching science content. Students were given time to plan experiments. They wrote predictions before beginning to experiment. For specific situations, they worked through the importance of controlling variables in designing experiments. They actively performed experiments. The lesson structure also stressed the importance of communicating results both orally and in writing.

In mathematics, teachers often followed a sequence of introducing the lesson's central concept by a problem-oriented whole class discussion followed by practice applying the concept either individually or in groups. Whole class discussions of solutions followed, leading to a lesson summary. The lessons were definitely teacher-guided, but with time for students to reflect on the lesson problem (and underlying concept) individually and in groups. There was a strong emphasis on problem-solving techniques. The approach also emphasized the idea that there is not one correct approach to a problem.

In one Population One class, the teacher asked students, "How can we get 17 on the calculator screen by using only the 5, +, -, /, and * keys?" Students worked individually with calculators to generate many key-press sequences that would produce the number "17" on the screen. These solutions were then presented to the class, discussed, and the solutions were reduced to succinct equations. The lesson concluded with the teacher writing on the blackboard, "How do we recognized 17?", followed by a number of different equations such as $17 = 85/5$, $17 = 22 - 5$, $17 = 15 + 2$, etc.

The importance of homework was not obvious from the observations. In mathematics lessons students were often asked to continue working on problems they had begun in class. Teachers also referred to supplemental books for students to use for additional problems. Science lessons often ended in the classroom but students were sometimes asked to conduct an observation or experiment at home or to write about something in their notebooks.

Lessons were obviously well-planned and organized by the teacher. The teacher was the classroom leader clearly directing all discussions. Many lessons began with teacher-led discussions which depended on student participation. Teachers used these discussions to bring out from students' suggestions the information needed to proceed with the lesson's concept. This approach seemed clearly to give students a sense of participation as they provided ideas from their own backgrounds and experiences.

A central problem facing teachers using this pedagogical strategy is whether or not they are engaging all students or only a select few. Active use of the "han" allows for student-to-student interaction often missing from other approaches. This use of groups helps solve the problem of engaging all students rather than a few. The presence of the "han" as an organized, continuing small group makes work in groups easier to handle and more efficient than it seems to be in many other countries. Children have regular experience working together with the members of their group. They learn about being responsible to and for each other. Since this use of groups is a consistent part of the characteristic pedagogical approaches used by most Japanese science and mathematics teachers in most grades, the experience of how to work responsibly and effectively in groups is cumulative and helps account for the comparative efficiency of the Japanese group work.

Some observations indicated that teachers were concerned when only a few students answered questions in discussions, especially those in many lessons' introductions. Teachers seemed to assume responsibility to motivate as many students as possible to participate in these discussions. Students were also given the opportunity to discuss their ideas together and later to present these ideas to the whole class for comments. A third type of interaction was also seen in Japanese lessons. The teacher interacted with individual students, usually when groups were working together on a problem or activity, or during an individual seatwork phase in the course of a mathematics lesson.

NORWAY

Across both mathematics and science, and both age groups, the most critical CPF aspects in Norway's observations were emphases on students' accumulating correct, factual knowledge and on students' involvement in exercises and learning activities. Students were expected to come to understand correct content through group and individual activities that were organized and supervised by the teacher. In general, teachers did not spend a significant amount of time lecturing or explaining concepts or topics. When teachers did present a lecture, their presentation was interspersed with questions to students both to elicit information and to focus students' attention on the topic being considered. New topics were introduced or continued topics further developed usually through a teacher-led class discussion, through reference to specific resource materials such as a film, textbook, or handout, or through involvement in a particular introductory or exploratory activity.

Much teaching in the Norwegian lessons depended on the students. Students were frequently questioned by the teacher. They were asked whether they understood something, such as homework exercises, something they had read, or ideas being discussed in class. They were asked whether they had completed an activity or assignment. Teachers used student responses as indicators for continuing a lesson. At Population Two, student-to-student or student-to-group questioning was often used rather than direct teacher inquiry to determine students' readiness to continue or to move to a new topic.

Teaching at both student levels was very student-centered. Further, in classroom activities, students were more explicitly involved with the subject matter than were teachers. This does not imply that teachers did not know their subject, were not thinking about the content, or were not carefully monitoring and evaluating classroom processes. It does fairly describe what was explicitly observable from what occurred in the classroom. Students were frequently called on to answer questions or perform activities. Even so, most students' responses were in written form. There was very little classroom-level subject matter discussion.

A fluid, but fairly consistent, lesson pattern was apparent in the Norwegian observations. Each lesson involved two major sections. The first section could include one of several elements but always involved a general introduction or transition into a lesson, a brief follow-up on previous activities or material, an introduction to a new topic, or an overview of a learning activity or experiment. The second section, usually the greater part of the lesson period, consisted of students working fairly independently on exercises, learning activities, or exploratory laboratory or hands-on exercises. Students often seemed free to choose whether they worked alone or in small groups.

Lesson content seemed neither complex nor structured on formal disciplinary lines. Lessons frequently dealt with basic facts and vocabulary. This included definitions or descriptions of simple concepts, such as respiration or boiling points, as well as descriptions of similarities and differences among groups of plants or animals. Many observed lessons were made of a range of barely-related or indiscernibly-related topics. This pattern was also found in the US. By contrast, this pattern especially surprised the French since such organization did not seem to permit thorough development of any one topic.

Use of textbooks and other resource materials was prominent in virtually all lessons. Students frequently worked alone or in groups on worksheets or on questions or exercises in textbooks. Teachers also relied heavily on the textbook not only for exercises and activities but also as an authoritative subject matter resource. Both teachers and students referred to textbooks during both the introduction and the development phase of a topic. Teachers almost seemed to avoid being considered subject matter experts and authorities — quite in contrast to the typical role taken by teachers in France. In one instance, the teacher, who did not know the answer to a question, left the classroom in order to look up the answer.

Few explanations were typically given in classroom discussion relevant to a lesson's focal topic. Teacher questions frequently concerned procedures rather than substantive content. Teacher comments similarly emphasized "what students must do" rather than "what they must understand". In mathematics and, especially in science, procedures were most emphasized by teachers.

Another indication of lessons' concerted student-orientation was the obvious effort teachers made to give non-evaluative responses to students' questions. They seemed to make great efforts to avoid negative comments. Teachers obviously wanted students to feel comfortable and, indeed, students seemed to do so. Teachers avoidance of criticizing a student's answer or telling students that they were not putting forth proper effort surprised the French SMSO members. The Japanese representative felt that it would not be an issue in Japanese classrooms. In France, such remarks would be made in order to remind students of their responsibilities and this was considered important. In Japan student effort was frequently encouraged. However, this was in a positive form exhorting students to work, rather than commenting on their failures to do so.

Perhaps as a result of this student-oriented, criticism-free approach, emphasis seemed to be placed almost entirely on correct students' answers. Only in a few instances were Norwegian teachers observed using students' mistakes to further a pedagogical aim. Correction mainly consisted in checking to be sure answers were correct with little if any discussion of the subject matter principles. It seemed that the teachers expected students to gain understanding through exercises, activities, and exploratory activities with, at most, responses from teachers indicating when they were correct. Correspondingly, students appeared to be considered to understand once they had given a correct answer. Teachers, especially at Population Two, rarely summarized or synthesized results obtained or knowledge acquired at the end of a lesson.

This student-orientation may also help explain why instructional goals for many learning activities and classroom practices seemed unclear. If one constantly relies on student accomplishments to provide direction for further work, providing "advanced organizers" or even clear outlines of topic development and the goals of activities related to that development could become problematic. Further, as data from Chapter Two made obvious, there is pressure to "cover" many more topics than expected in other countries other than the United States. Teachers who feel the need to cover many topics and to actively involve students in all aspects of their learning would likely structure classroom activities around student learning and devote considerable time to individual and group-oriented exercises and experiences. With next stages almost always contingent on student response to current activities, clear aims are more difficult to state and may well be considered inappropriate for a student-centered approach.

SPAIN

The CPF of observed lessons in Spain appeared in many respects quite similar to that from France. The most characteristic pedagogical issues again involve teachers and lesson content. In the observed lessons, the teacher usually introduced students to formal rules, definitions and procedures of subject matter topics. They also presented specific applications or illustrations and then gave students time for individual practice and for applying what had been presented. The teachers in Spain, as were those in France, were actively involved in presenting content to students.

Content complexity in the observed lessons ranged from content that would be considered challenging in other countries, such as verifying Euler's theorem, to content that would typically be considered review for Population Two students — such as addition, subtraction, and multiplication of positive and negative whole numbers, and properties of multiplication. Nonetheless, topic presentation in both mathematics and science appeared focused on the subject matter discipline's structure. For example, one science lesson discussed a homework project and included a review of Pascal's law on the pressure of a fluid in a closed system. In another lesson discussion focused on the principles of dominant and recessive traits in genetic inheritance.

In most lessons, presentation of a principle was immediately followed by some consideration of its practical implications or applications. Teachers made frequent use of the blackboard, photocopies, and graphics in the textbooks to provide visual representations of principles either in iconic form — such as a graph — or through practical, everyday illustrations. Indeed, a striking feature of Spanish lessons was teachers' consistent linking of principles studied with practical, everyday applications. Teachers seemed to make concerted efforts to link current topics to students' prior learning, general experience, and knowledge of everyday life.

Both mathematics and science lessons shared a fairly common structure. Lessons began either with a review of assigned homework or a review of the previous lesson and of principles from earlier lessons important for the present day's topic. Apparently, from the considerable time devoted to it, homework played an important role in lessons. Indeed, mathematics lessons in particular seemed almost to be organized around homework. New topics were not introduced or continued topics developed further until the teacher was satisfied that students had sufficient understanding. This sufficiency was judged largely by homework and discussions of it.

Following the lesson's review phase, new topics or further topic development took place. This frequently involved use of the textbook. For example, in a science lesson one student read to the class from the textbook while in a mathematics class the teacher read from the textbook as students followed along. In another mathematics class, the students read the relevant portion of the textbook to themselves. Beyond textbook use, teachers frequently asked the whole class questions to which individual students responded. These questions were often integrated into the topic material presentation. Their purpose was apparently both to assess students' understanding and comprehension and to increase students' involvement with the subject matter.

Teachers followed topic presentation with time for students to practice or apply what had just been covered. Presentation and practice phases often alternated more than once in the same lesson, in a pattern similar to that found in France. Lessons often concluded with teachers assigning homework or in some other way identifying the topic of the next lesson for students.

The obvious and integral use of textbooks by both teachers and students suggested that, unlike the French lessons in which teachers were clearly the subject matter experts, in Spain subject matter was organized and actively presented by the teacher but mediated by the textbook. In almost every lesson, homework was either assigned, corrected, or both. The review and correction of homework constituted a sizable portion of some lessons. However, unlike lessons in France but similar to lessons in Norway and the US, students also had considerable lesson time in which they worked alone on exercises. The use of this seatwork time was more integral to the lesson's structure and more a part of overall interaction, unlike similar seatwork in Norway and the US.

SWITZERLAND

All lesson's observed in Switzerland were made in one German Canton. The significance of this cannot be appreciated unless one reads the Swiss Case Study presented in Part II of this book or is already aware that Swiss education differs considerably by regions having different primary languages of instruction and culture: German, French, and Italian.

More than any other country in which observations were made at both student population levels, the German Swiss lessons showed significant CPF differences between Population One and Population Two. In general, across all lessons, emphasis was placed on students responsibility to learn (Beck, Guldimann, Zutavern, 1991; Deci & Ryan, 1993). Students were expected to explore and come to understand content by completing learning activities or observing demonstrations organized and directed by the teacher. Several lessons at both population levels made explicit mention of a "weekly plan" which outlined students' learning goals and activities. Students were then expected to organize and plan their use of time both in the classroom and outside it to accomplish the specific learning goals and tasks. This was assigned a high priority in Switzerland to encourage development of each individual student's sense of responsibility for learning (EDK, 1995; NW EDK, 1995; Ramsegger, 1993; Simons, 1992).

Particularly in Population One, lessons seemed almost to be organized more around students' active involvement in — and responsibility for — their own learning, rather than around specific content goals. For example, in one situation the teacher varied the quantity and type of exercises for each student. All students were expected to complete a common minimum content. Upon completing this common core, the teacher gave the students a test. If they passed, they could choose between four different types of additional exercises that were also included in the weekly plan. Thus, all students worked on the same content but not all of them completed the same learning activities or exercises.

Topic complexity was another unique CPF aspect in Population One lessons. Science lessons topics for these students' seemed rather basic compared to topics for similar students in Spain and the US. Teachers seemed particularly to choose topics that would appeal to students' interest and motivate their involvement. For example, teachers told stories about animals or prehistoric people, asking students for hypotheses and ideas that students then explored through discussion and finally drew some conclusions with the teacher's guidance. In contrast, the content of these younger students' mathematics lessons exhibited a surprising range of complexity from simple addition, subtraction, multiplication, and division to the use of Venn diagrams, discussion of sets, set member inclusion, and set intersection.

Perhaps because of this strong emphasis on student-centered instruction, no typical lesson structure emerged from Population One lessons. By contrast, despite a similar student-centered emphasis at Population Two, lessons followed a fairly consistent pattern. Lessons had two major phases in both mathematics and science. First, teachers introduced a topic. This introduction might involve explaining or giving an overview of a learning activity, a discussion in which the teacher asks students questions about a topic, or a demonstration by the teacher of a particular procedure or principle.

Lessons typically were structured around a single subject matter principal or concept. However, teachers lectured very little. Each lesson appeared to cover a small but feasible amount of content. Teachers showed little obvious concern about covering a specific quantity of content. Rather, the process seemed very inductive beginning by presenting some exercises or examples from which the topic was then developed. Teachers organized learning activities, presented demonstrations and asked students questions. Students were often asked questions such as "What is the meaning of the term ...?", "What do you observe?," or "What happens when...?" to encourage reflection and reasoning about the topic under consideration.

In the second part of these lessons, students had the opportunity to work on a learning exercise or complete an exploratory laboratory activity following the teacher's demonstration. Often students worked alone on worksheets, exercises, or in writing down their observations from the teachers observations. Teachers frequently allowed students to work together on these activities, although many if not most students worked alone.

Unlike Spain, where teachers made obvious use of textbooks, or Japan, where similarity between different teachers' lessons suggests an underlying reliance on some common documents, teachers in the Swiss lessons showed no dependence on particular curricular or pedagogical resources. The predominant emphasis was on students' responsibility to learn. In fact, the comparatively low profile of Swiss teachers prompted one French observer to comment that "teachers do not teach." Teachers and books served as resources for students in their learning activities but lessons were structured around student learning activities and teacher demonstrations. Although teachers may not appear active in this approach, structuring students' learning around exploration and specific learning activities requires considerable prior thought, planning and organization by teachers. The lesson structure is precisely what one would expect where the necessity for students to take responsibility for their own learning and to create meaningful understanding for themselves was the dominant understanding of education (Beck, Guldimann, Zutavern, 1991; Edelstein, 1992; EDK, 1995; Klippert, 1993; NW EDK, 1995; Ramsegger, 1993, Simons, 1992; Thomas, 1992).

UNITED STATES

The role of teachers was central to the characteristic pedagogy in the US. In general, teachers acted as information transmitters. Teaching appeared designed to facilitate learning by providing needed information. Teachers seemed more involved with subject matter content than did students although, in some lessons, very little actual topic content was observable. Teachers clearly took primary responsibility for the learning process.

While individual lessons may have been highly structured, the pattern across lessons and teachers was of considerable diversity. Lessons were structured, organized and directed by the teacher often with little, if any, obvious influence from the students. Lessons were structured with either a content focus or a learning activity focus. Content-centered lessons typically had three parts: an information section, an elicitation section, and an individual practice section. During the information section, teachers introduced the lessons' content, perhaps with the aid of the blackboard or overhead projector. Teachers explained complex concepts to students by giving definitions, offering concrete examples and providing demonstrations. These explanations and development of new concepts frequently were on a rather formal, abstract level. Teachers then checked students' understanding during a question-and-answer period. In contrast to teachers' presentations, questions asked students were less abstract and more content- or product-oriented, e.g., asking for definitions or the answers to specific exercises. During the third part, students were given the opportunity to practice what had just been presented. Students worked individually on exercises with the teacher often helping students that had difficulty. Students were usually told that their work would be checked at the end of the period.

Lessons with an activity focus had two phases: a brief introduction or explanation of the activity by the teacher and a time in which students worked on the activity, problem or experiment. In this type of lesson, students frequently worked in groups to complete the experiment or activity.

Topic complexity for Population One and Two lessons seemed little different from that in Norway or Switzerland. However the theoretical and abstract language used in one Population One science lesson — "energy," "proton," "neutron," "electron," and "atom" — prompted several representatives to question the appropriateness of such terminology. Subject matter content in general appeared to be represented in more theoretical, abstract and procedural form rather than represented in practical situations or related to students' experience. This seemed particularly true for younger students.

Providing definitions of ideas, words or concepts as they are written in a dictionary was emphasized. In one instance, students were to copy and memorize definitions from a dictionary. Consistent with this emphasis on written information, written homework was a common feature. However, little was done with students' homework other than to check it for correct responses or, in some instances, merely to check whether the assignment had been completed or not. The almost cursory attention given homework during most of these lessons left its relationship to the substance of the lesson a mystery. It always seemed important to have completed the homework but it was not clear what purpose the homework served or even if students had actually learned something by doing it. Only rarely did homework serve as a platform for the current lesson's work; rather, it seemed at best closure on the work of the previous lesson and a way of securing out-of-class study from students.

In the US, content review did not appear as a separate, identifiable part of the overall structure of lessons. One explanation for this may be the amount of redundancy built into the curriculum across grade levels as portrayed in Chapter Two. Review may be regarded as something done more in large units early in a school year to re-introduce content from the previous year — an approach that may be the function of schooling in a highly mobile society. This content redundancy is not a new finding. Comparing three mathematics textbook series covering grades K-8, Flanders (1987) found from 35% to 60% of the textbook material in the texts for grades 2-5 was review. Review accounted for about 55% of the material in fourth grade texts and as much as 70% of the material in eighth grade texts (fourth and eighth being the two US focal grades here). Data in Chapter Two suggests that the situation is likely quite similar in science. With so much redundancy and review built into the curriculum, teachers may not feel the need to review concepts daily or may depend on the structure of the textbook to ensure that all students have gained conceptual understanding of topics needed. If not, of course, they will encounter these topics again and have yet another opportunity to learn.

DISCUSSION AND CONCLUSIONS

What guides teachers activities in a classroom? What assumptions do they make? What understanding of content and pedagogy do they have? How do they act to unfold content during the course of a lesson? What strategies do they use? Are the pedagogical and lesson-structuring practices of teachers in similar settings also so similar that broad characterizations of pedagogy are possible? This chapter has sought an empirical investigation of these questions, based largely on observational data and the case studies reported later in Part II of this volume.

"Characteristic pedagogical flow" (CPF) was presented as an organizing motif for how teachers interactions in classrooms are structured. For experienced teachers in typical classroom situations with reasonable resources, teaching is a flow activity. Reflection, analysis and deliberate decision making take place before or after a lesson's activity, but routine and practice largely govern behavior during lessons. In this gestalt behavior, beliefs about subject matter, presenting content, what learning is, the role of a teacher, how students learn, and so on, is woven into a whole that reflects certain beliefs and rationales even when teachers are not aware of them while actually teaching.

The contention here has been that this pedagogical flow is sufficiently shaped by common training and experience that, for groups of similar teachers (for example, those at a specific grade level teaching a specific subject in one country), this pedagogical flow of teacher behavior in the classroom has characteristic features that can reveal important things about an entire group of teachers — characteristic beliefs, values, practices, and so on. This chapter has focused on three potential characteristic features: (1) content representation and complexity, (2) content presentation in lessons, and (3) the nature of classroom interactions and discourse, particularly, their relation to content. CPF has been presented in this chapter as an organizing principle informing and guiding teacher action as curricular intentions are shaped through classroom practices.

The CPF conception is consistent with research that has emphasized the importance of subject matter content and the way in which teachers think about this content in instruction (Stodolsky, 1988; Clark, 1986; Shulman, 1986a, Thompson, 1992). In particular, although conceptual, the CPF notion stems from an empirically-driven attempt to explain qualitatively different but characteristic lessons observed in six countries. The differences observed did not appear to reflect simply maps pairing subject matter topics with instructional techniques or approaches. Rather deeper aspects of subject matter content, as it is reflected in teachers' thinking and in their curricular intentions, interacted with various instructional practices and the beliefs underlying them to yield qualitatively different types of practices, many aspects of which could be observed in classroom lessons and used as a basis for characterizing and explaining pedagogical practice for mathematics and science.

CPF seems a powerful, general notion. For the empirical data available, only a modest beginning investigation of CPF was possible. Three key aspects of CPF — and the beliefs and assumptions underlying them — were examined here. First, content complexity and representation ("What is the lesson's content?") was examined. What constitutes subject matter, whether particular topics are included or not and the manner in which included topics are developed in the curriculum, results in qualitatively different experiences of subject matter for students. In addition, the understandings and perspectives teachers have about subject matter influence their choice and implementation of activities which further shape and shade students' learning experiences.

Second, content presentation ("How is the content presented?") was investigated. Instructional activities differ such that some enable students to develop better understanding than others. The choices of how to present or unfold content in a lesson involves deeply ingrained teacher beliefs about teaching, learning, pedagogy, subject matter, and so on. How instructional activities are incorporated into lessons and contribute to topic development and understanding is a major factor in producing characteristic qualitative differences observed across lessons.

Third, content discourse ("How do students and teachers interact with each other and the lesson content?") was discussed. Communication is fundamental in learning. The questions teachers ask students, the choice of analogies and examples, and the manner in which explanations are developed and presented all have an impact on the character of a lesson and the quality of students' learning. The discussion earlier revealed characteristic differences among SMSO countries for various aspects of content discourse.

The basic findings of this chapter are simple but fundamental. The important variation among lessons probably is not quantitative but qualitative. Lessons do differ in important ways along the key dimensions of CPF. Characteristic national practices can be identified — and seem so fundamental that, not only are they qualitatively different, but, with sensitive instruments to gather appropriate data, inter-country variation may well outweigh intra-country variation. These issues are especially significant in comparative analyses of students' learning across countries — and, in particular, for providing contextual and explanatory background for TIMSS.

The CPF notion may be viewed as an extended hypothesis. On the basis of data discussed here and in Chapter Two, we have concluded that there are nationally distinct CPF perspectives and that the differences observed in student performance across countries can only be understood and interpreted through a consideration of CPF. The work reported here suggests that some consideration of CPF is essential to the interpretation and analysis of comparative educational studies. Its value within TIMSS remains to be determined. Its potential for comparative studies and for understanding pedagogical practice more generally has barely been touched.

Chapter 4:
Moving from Conceptions to Instrumentation

The central charge of the SMSO project, no matter what additional results were obtained, was to develop instruments for collecting data in TIMSS. This included developing questionnaires for teachers and students. The observations, logs and case studies previously discussed aimed at uncovering qualitative differences in lessons, rather than more surface descriptive features.

SMSO made concerted efforts to incorporate into the survey instruments developed the issues and concepts capturing these important qualitative differences. The classroom observations strongly suggest CPF issues must be kept in mind in reporting and interpreting cross-national results. This type of information is essential if cross-national comparative research's purpose is to learn something of the relationship between schooling and achievement. Many traditional, simple indicators of instructional practices are not sufficient to capture the rich variation noted in the observations. In addition, the observational work suggests that causal inferences that may be drawn among a collection of simple indicators may differ from one country to another, e.g., the relationship between amount of homework and students' achievement. Cross-national comparative survey research may need to be supplemented by non-survey methods to capture the truly important issues that distinguish countries.

The classroom observations prompted a fundamental reorientation in the instrument development. Original development began with more traditional indicators, especially those relevant to a US context. The final TIMSS' instruments were conceptually designed as an attempt to capture some salient aspects of CPF. The CPF concept suggests that what constitutes subject matter is important to the understanding of both teaching and learning. If subject matter importance is taken seriously, some measure of subject matter content is essential. Subject matter's importance has often been invoked in comparative education discussions but efforts to measure subject matter or curriculum have frequently not been included in designs, analyses and considerations — especially in large scale cross-national studies. The CPF notion also involves the interaction of subject matter content and pedagogy in creating lessons of a particular character. The TIMSS curriculum frameworks were a concerted effort to describe curriculum with some richness. They allowed portrayal of several types of curricular complexity: topic, developmental, and cognitive. The frameworks' content aspect was employed to describe curricular topic and developmental complexity while the performance expectation aspect described information relevant to curricular cognitive complexity. The frameworks provided a common language system that made possible multidimensional examination of all aspects of curriculum — intended, implemented and attained. Based on this common language, the research instruments and procedures for TIMSS contained measurements of curriculum for all three aspects making possible, perhaps for the first time, consistent measurement and comparison across what is intended to be taught, what is potentially taught (e.g., textbooks), what is actually taught in classrooms, and what students have learned (see Figure 1-2, in Chapter One).

The curriculum analysis procedures investigated the intended curriculum by providing a multidimensional portrait of curriculum documents — e.g., guides and textbooks — that inform, guide, and otherwise influence classroom lessons in mathematics and the sciences. The in-depth document analysis employed low inference procedures carried out by trained subject-matter experts in each country to portray curricular documents by identifying the specific subject matter topics included in a curriculum (content complexity) and what students were expected to do with these topics (cognitive complexity). Two other analysis procedures for curricular intentions, general topic trace mapping (GTTM) and in-depth topic trace mapping (ITTM), made possible a broad look at how curricular topics were intended to be sequenced and organized across the years of schooling (developmental complexity). These procedures together made possible a multidimensional portrait of National/Regional Curriculum Goals, one aspect of the Provision of Educational Experiences model outlined in Chapter One, and also helped provide a partial response to the question, "What are students expected to learn?" (see Figure 1-2, Chapter One). This multifaceted description provided especially important perspectives on textbooks, one of the most important resources informing and guiding teachers' classroom instruction.

INVESTIGATING CONCEPTS AT THE IMPLEMENTED LEVEL: ILLUSTRATIONS FROM ONE APPROACH

The SMSO project tried to develop instruments that addressed subject matter content and pedagogy, important aspects of CPF coming into play as curricular intentions and aims were pursued in classroom lessons (implemented curriculum). The TIMSS Teacher Questionnaire was the only instrument for gathering information directly from teachers. It became the central tool with which to measure aspects of curricular complexity and pedagogy as these are practiced in classrooms. A brief description of the Teacher Questionnaire follows, together with small portions of some specific items developed in the attempt to assess both of these important aspects of CPF. While all items included in the Teacher Questionnaire were conceptually linked to the CPF notion through the constructs identified in the model of Educational Experiences presented in Chapter One, the particular items presented here represent those most relevant to CPF as it could be investigated by survey methods. A more exhaustive consideration of all the items and instruments is available elsewhere (see SMSO, 1993c).

What is presented represents only one approach to measuring concepts and issues important to CPF. No claim is made for their sufficiency in creating a clear, complete, and satisfying portrait of CPF. They do represent a serious start at measuring some CPF aspects, subject to the constraints of survey methods. Other approaches are both possible and necessary to portray fully and completely the complex interaction between subject matter content and pedagogy revealed as so obviously important by the classroom observations. Given these limits, the items that follow represent examples of products of the discourse methodology outlined in Chapter One and have demonstrated, in informal and formal pilot investigations, considerable potential for assessing important aspects of curriculum and pedagogy across countries. Their ultimate test awaits the TIMSS achievement results.

ASPECTS OF CPF: SUBJECT MATTER CONTENT

Items to describe Teachers' Content Goals (see Figure 1-3, Chapter One) were designed to identify and characterize (according to the appropriate TIMSS curriculum framework) the subject matter content on which teachers focused in their classrooms. The items developed for this purpose include an assessment of teachers' content goals, teachers' indication of students' opportunity to learn, and teachers' indication of what they consider to be on ideal student response to selected specific exercises.

CONTENT GOALS

In the Content Goals item, teachers are to indicate the extent to which they address specific topics in lessons over the school year and whether these topics have been addressed in the instruction of previous years. This identifies those topics teachers actually present in the classroom (i.e., topic complexity) and gives some indication of the sequence of curricular topics relative to the current year (i.e., developmental complexity). The topics listed represent the complete spectrum of the relevant framework, either mathematics or the sciences. These are measured at a more global level of the frameworks than was used in the curriculum analysis. Descriptions of framework categories were recast to be more accessible to teachers. The TIMSS in-depth topics, topics of special interest and focus in curriculum analysis and testing, are also included here separated from the category that might otherwise contain them. This measurement of the implemented curriculum enables direct links to the intended curriculum through the TIMSS frameworks. An example of how these concepts were implemented can be seen from the small portion of the Content Goals item for Population Two mathematics teachers displayed in Figure 4-1.

7 *continued*

TOPIC AREA	HOW LONG DID YOU SPEND TEACHING EACH OF THESE TOPIC AREAS TO THE <TARGET CLASS> THIS YEAR? WILL YOU COVER ANY OF THESE TOPICS IN FUTURE <PERIODS>?						
	have covered this year <periods> completed				*will cover later this year*	*not covered this year*	COVEREDA PREVIOUS YEAR
	1-5	6-10	11-15	> 15			
1. Whole Numbers	☐	☐	☐	☐	☐	☐	☐
Indicate your coverage both at the main topic level and for each of the following subtopics.							
• Meaning of whole numbers; place value and numeration	☐	☐	☐	☐	☐	☐	☐
• Operations with and properties of whole numbers	☐	☐	☐	☐	☐	☐	☐
2. Common & Decimal Fractions	☐	☐	☐	☐	☐	☐	☐

Check as many boxes as apply for each topic listed.

Figure 4-1. Teacher's Content Goals

OPPORTUNITY TO LEARN (OTL)

Two additional items were developed to help measure curriculum as embodied in classroom lessons: the Opportunity-to-Learn (OTL) section and the Ideal Student Response item. Both were developed around in-depth topics. The OTL section provides additional information about topic and developmental complexity for these specific topics. Using items from the student achievement test pool to illustrate each in-depth topic, teachers are asked to indicate when students have studied the topic. Teachers are also asked to indicate whether they consider the items appropriate for assessing that particular topic with their students. As each assessment item was coded for performance expectations as well as content, this section provides some data on teachers' conceptions of the cognitive complexity of the topics, as well as indications of the teacher's assessment expectations (and how these match assessment through the TIMSS test). Concentrated comparison and consideration of these in-depth topics is thus made possible across curriculum levels — analysis of curriculum documents, teachers' indication of students' Opportunity-to-Learn, and measurement of students' achievement. Figure 4-2 contains an example of this item type from the Population Two Science Teacher Questionnaire.

12.   **LIGHT**

The following exercises illustrate this topic. These exercises, or ones like them, might be used to assess students' learning of this topic.

i.   The walls of a building are to be painted to reflect as much light as possible. What color should they be painted?

ii. A flashlight close to a wall produces a small circle of light compared to the circle it makes when the flashlight is far from the wall. The same amount of light energy reaches the wall regardless of distance. Explain why.

iii. A person in a dark room looking through a window can clearly see a person outside in the daylight. But a person outside cannot see the person inside. Why does this happen?

iv. A beam of light strikes a mirror as shown. What picture would best show what the reflected light would look like?

A.   Is anything done in your science class that would enable your students to complete similar exercises that address this topic?
     Check as many as apply.

     YES...
     i.   Something was done EARLIER this year. .... ☐
     ii. Something is CURRENTLY in progress. .... ☐
     iii. Something will be done LATER this year. .... ☐

     NO...
     iv.   The topic was covered in the curriculum for an EARLIER grade. ....☐
     v.   Although the topic is in the curriculum for THIS grade, I will not cover it. ..☐
     vi.   The topic is covered in the curriculum for a LATER grade. .... ☐
     vii. To my knowledge, this topic is NOT INCLUDED in the curriculum. ....☐
     viii. I DO NOT KNOW whether this topic is covered in any other grade.   ....☐

B.   If you were to develop a test for your target class that assesses this particular science topic, which of the above items would you consider appropriate for the test?
     Check all that apply.

     ☐ i   ☐ ii   ☐ iii   ☐ iv   ☐ none

C.   Are students likely to encounter this topic   <u>outside</u> of school this year?
     Check one.

     ☐ Yes   ☐ No

Figure 4-2. Opportunity to Learn (OTL)

IDEAL STUDENT RESPONSE

Content representation can be portrayed by characterizing teachers' indications of ideal student responses to specific items. This provides a critical link between content complexity and pedagogy. Identifying the topics addressed in classroom lessons (Teacher's Content Goals and OTL) is only part of the story. Knowing how topics are represented and implemented is also essential in understanding students' learning and attainments. Topics may be introduced and developed with students from varied subject matter perspectives and expectations for students. In this survey item approach, teachers provide what they consider to be an ideal student response to an actual extended response (longer, open-ended) item from the TIMSS test. The exercises to which teachers respond are to be taken directly from the in-depth topics as assessed in the student achievement instruments. This section was included in the field trial version of the teacher questionnaire and the results were promising. Due to political and practical considerations (especially teacher response time, and the cost and complexity of coding open-ended questionnaire items) the item was eliminated from the final TIMSS Teacher Questionnaire. The richness and variety of teachers' responses can be seen in Figure 3 which contains two examples from two of the SMSO countries.

The types of solutions teachers use in teaching students how to solve specific types of problems are very important. **WHAT TYPE OF SOLUTION WOULD YOU SUGGEST TO STUDENTS AS ONE OF THE BETTER WAYS TO SOLVE EACH OF THE PROBLEMS BELOW?**

Please detail all the steps you would expect to see on a student paper.

A building that is 24 m high casts a shadow of 18 m. At the same time, a flagpole casts a shadow of 15 m. How high is the flagpole?

Solution Steps

Teacher A's Response

Teacher B's Response

We can represent the situation with a figure.

$$\left.\begin{array}{l}(AB)\perp(AC)\\(A'B')\perp(AC)\end{array}\right\} \text{ therefore } (AB) \, // \, (A'B')$$

In the triangle CAB

$$\left.\begin{array}{l}CA' \text{ and } A\\CB' \text{ and } B\end{array}\right\} \text{ are aligned in this order}$$

and $(AB) \, // \, (A'B')$

therefore, according to Thales' theorem

$$\frac{CA'}{CA} = \frac{A'B'}{AB}$$

therefore, according to the hypothesis

$$\frac{15}{18} = \frac{A'B'}{26} \text{ and } A'B' = 20$$

The flagpole measures 20 meters.

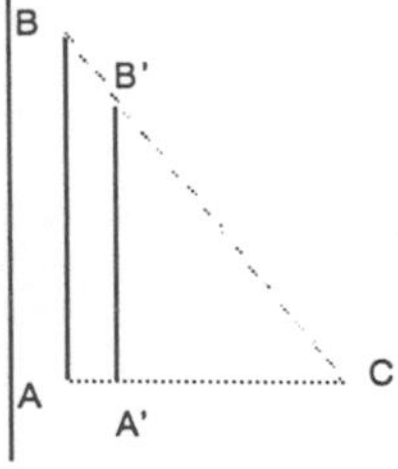

1. $\dfrac{24}{18} = \dfrac{x}{15}$

2. $18x = 360$

3. $\dfrac{18x}{360}$

4. $x = 20$

Figure 4-3. Ideal Student Response

ASPECTS OF CPF:
PEDAGOGY AND INSTRUCTIONAL PRACTICES

Several items were also developed to address the pedagogical aspect of CPF. Some of the important notions relative to this pedagogical aspect are how content is represented, presented and discussed during lessons. All of these items are related to the Instructional Practices concept in the Provision of Educational Experiences model presented in Chapter One. In each case, items were designed to address one of the features found to characterize important CPF differences encountered in the different countries' classroom observations as described in Chapter Three. The features found central in these SMSO cross-national portraits motivated the development, form and content of specific items for the TIMSS Teacher Questionnaire. Among these items were the teacher's report of a recent lesson's structure; the students' report of the frequency with which various instructional practices occur in lessons (on the TIMSS Student Questionnaire); a characterization of the nature, frequency and use of lesson-related homework; and an indication of the type of discussion about lesson topics teachers expected from students during lessons.

LESSON STRUCTURE

How content is presented — the structure and coherence of lessons (see Figure 1-3, Chapter One) — was explored in the teacher questionnaire through a detailed profile of a specific lesson. From the observational work reported in Chapter Three, the important differences among lessons were found not so much in the type of activities included but in the manner in which they were incorporated into the overall lesson development or "flow". The amount of time devoted to particular lesson elements, such as homework review and individual student practice time, and the way these elements were sequenced within the lesson contributed to the lesson's character. The "Structure of the Lesson" item was developed as an attempt to capture and assess these obvious qualitative differences in lesson organization and structure observed by the SMSO team. The result of this effort is presented in Figure 4-4.

THINK OF THE SAME MATHEMATICS CLASS <HOUR/PERIOD>.

A. HOW DID THE LESSON PROCEED? The following presents a list of activities that may occur during a lesson. Although the list is not exhaustive of what is done in a classroom, most activities may be considered as variations of those listed below. Using this list, indicate how your lesson developed. In the blanks, write in the order in which the activities you used in the lesson took place (1 = first, 2 = second, and so on) and estimate the amount of time you spent in each one. Ignore activities you used that do not fit into the descriptions listed. _Write in the order and the approximate number of minutes for each activity that you used_ — _leave blank any activities you did not use_. NOTE: If you used the same activity more than once in the lesson, write in the order and minutes for each time you used it.

	order	minutes
• review of previous lesson(s)	_____	_____
• a short quiz or test to review previous lesson	_____	_____
• oral recitation or drill (students responding aloud)	_____	_____
• review or correction of previous lesson's homework	_____	_____
• **introduction** of a topic (class discussion, teacher explanation/demonstration, film, video, use of concrete materials etc.)	_____	_____
• **development** of a topic (class discussion, teacher explanation/demonstration, group problem solving, film, video, etc.)	_____	_____
• small group activities (with or without teacher)	_____	_____
• students do paper-and-pencil exercises related to topic (not the same as homework)	_____	_____
• assignment of student homework	_____	_____
• students work on homework in class	_____	_____
• student laboratory or data collection activity (not a separate laboratory hour) or hands-on session	_____	_____

Figure 4-4. Structure of the Lesson

TEACHER AND STUDENT INTERACTIONS

How teachers and students interact with topics and each other during lessons is another key aspect of CPF (see Figure 1-3, Chapter One). What teachers expect students to do as assessed by the performance expectation aspect of the curriculum frameworks is one way to approach this issue of cognitive complexity. Another is to assess how teachers involve students in discussions around lesson topics. Figure 4-5 presents one approach to measuring this from the Science Teacher Questionnaire. The important focus in analyses of this item will not be the relative frequencies for any particular option but the pattern of choices teachers make across the entire set of options.

IN YOUR SCIENCE LESSONS...

A. HOW OFTEN DO YOU TYPICALLY ASK STUDENTS TO DO THE FOLLOWING?

	never	rarely	some-times	always
1. Explain the reasoning behind an idea	☐	☐	☐	☐
2. Represent and analyze relationships using tables, charts, or graphs	☐	☐	☐	☐
3. Work on problems for which there is no immediately obvious method of solution	☐	☐	☐	☐
4. Use computers to solve exercises or problems	☐	☐	☐	☐
5. Write explanations about what was observed and why it happened	☐	☐	☐	☐
6. Put events or objects in order and give a reason for the organization	☐	☐	☐	☐

Figure 4-5. Lesson Discourse

An interesting difference found in the observations involved another aspect of classroom interactions, specifically, how teachers responded to student misconceptions or inaccurate responses. An item was developed to measure these differences and asks, "In your mathematics lessons, how frequently do you do the following when a student gives an incorrect response during a class discussion?" Respondents are asked to check a box after each option to indicate whether it is something they do "never or almost never", "some lessons", "most lessons", or "every lesson". A total of four options are listed including "correct the student's error in front of the class", "ask the student another question to help him or her get the correct response" and "call on other students to get their responses and then discuss what is correct." A parallel item was developed for science.

HOMEWORK

As described in previous chapters, homework was one instructional activity that contributed significantly to the different character of lessons (see Figure 1-3, Chapter One). The obvious, qualitative differences discussed in Chapter Three, however, were not due to the amount of time students were expected to spend outside of class doing their homework. Time spent on homework has been the main focus in many other studies and is something that would not be obvious from classroom observations. Rather, the observed differences related to the type of homework students did and the way teachers incorporated homework into subsequent lessons and topic development.

Figure 4-6 contains the science version of the items developed to capture these qualitative differences. Again, the important focus in analyses of this item will not be the relative frequency a teacher indicates for any particular option but the pattern of choices teachers make across the entire set.

SCIENCE HOMEWORK

A. DO YOU ASSIGN HOMEWORK? *Check one box.*

☐ NO   ☐ YES

B. IF YES, HOW OFTEN DO YOU TYPICALLY ASSIGN THE FOLLOWING KINDS OF SCIENCE HOMEWORK? *Check one box in each line..*

	never	rarely	some-times	always
1. worksheets or workbook	☐	☐	☐	☐
2. problem/question sets in textbook	☐	☐	☐	☐
3. reading in a textbook or supplementary materials	☐	☐	☐	☐
4. writing definitions or other short writing assignment	☐	☐	☐	☐
5. small investigation(s) or gathering data	☐	☐	☐	☐
6. working individually on long term projects or experiments	☐	☐	☐	☐
7. working as a small group on long term projects or experiments	☐	☐	☐	☐
8. finding one or more uses of the content covered	☐	☐	☐	☐
9. preparing oral reports either individually or as a small group	☐	☐	☐	☐
10. keeping a journal	☐	☐	☐	☐

C. WHEN STUDENTS ARE ASSIGNED <u>WRITTEN</u> SCIENCE HOMEWORK, HOW OFTEN DO YOU DO THE FOLLOWING? *Check one box in each line.*

	never	rarely	some-times	always
1. only record whether or not the homework was completed	☐	☐	☐	☐
2. collect, correct and keep assignments	☐	☐	☐	☐
3. collect, correct assignments and then return to students	☐	☐	☐	☐
4. give feedback on homework to whole class	☐	☐	☐	☐
5. have students correct their own assignments in class	☐	☐	☐	☐
6. have students exchange assignments and correct them in class	☐	☐	☐	☐
7. use it as a basis for class discussion	☐	☐	☐	☐
8. use it to contribute towards students' grades or marks	☐	☐	☐	☐

Figure 4-6. Homework

OTHER ASPECTS OF INSTRUCTION

In addition to the Teacher Questionnaire, several items were included in a Student Questionnaire to provide additional data on typical instructional practices in lessons. Each of these items is presented with specific reference to mathematics, science, or to a specific science for older students (e.g., biology, chemistry, earth science or physics). For example, one item asks, "How often does this happen in your science lessons?" This is followed by a list of 16 different classroom activities such as "We copy notes from the board", "The teacher gives a demonstration of an experiment", "We discuss our completed homework", "We work together in pairs or small groups", and so on. A second item asks, "When we begin a new topic in science, we begin by...." which is followed by seven different activities. These include activities such as "having the teacher explain the rules and definitions", "working together in small groups on a problem or project", and "trying to solve an example related to the new topic." For both of these items, students are to choose whether each activity occurs "almost always", "pretty often", "once in a while", or "never."

The type of and manner in which teachers use and incorporate various resources into a lesson also contributed to its character. Several items on the teacher questionnaires relate to one key resource, textbooks (see Figure 1-3, Chapter One), and provide a vital link between measurements of the intended and implemented curriculum. Teachers are asked to indicate what textbook, if any, they use in teaching their mathematics and science classes. Teachers may indicate which textbook they use by selecting from a list of the textbooks analyzed in the curriculum analysis project in their country or by filling in the textbook title and publisher information. Teachers are also to indicate the approximate percentage of their weekly teaching that is based on the text they've indicated. They may choose one of four responses, "0-25%", "26-50%", "51-75%" or "76-100%."

How teachers assess students' learning and the use they make of these assessments contribute to the qualitatively different school experiences for students (see Figure 1-3, Chapter One). Teachers are asked to indicate how much weight they give seven different assessment methods in assessing students' work. The assessment methods include "teacher-made short answer or essay tests that require students to describe or explain their reasoning", "teacher-made multiple choice, true-false and matching tests", and "how well student do on projects or practical/laboratory exercises." Teachers indicate the weight they give each type by selecting "none", "little", "quite a lot" or "a great deal." A second item asks, "How often do you use the assessment information you gather from students to...". This is followed by six purposes including "provide students' grades or marks", "diagnose students' learning problems", "report to parents", and "plan for future lessons." Teachers indicate their response for each purpose by selecting from the same set of four responses presented for the previous item.

OTHER FACTORS INFLUENCING CPF: TEACHER'S BACKGROUND AND CONTEXT

The CPF notion emerged through discussion centered around classroom observations. It was clear that subject matter and pedagogy were essential aspects of this notion. The fact that there were such qualitative distinctions across countries suggested that systemic factors contributed to these differences and needed to be measured and understood. Classroom observations permitted only descriptions of what was apparent. The CPF notion was developed in seeking an "explanation" of the observed differences. Through interviews with teachers and the discourse approach, hypotheses were entertained as possible explanations for teachers' organization and discourse during lessons. In addition to the instructional practices already described, systemic factors of background and beliefs needed to be measured to more fully understand CPF.

A number of characteristics of the system and school in which instruction occurs as well as characteristics of the teacher who delivers the instruction were considered to influence the character of lessons. This led to the inclusion in the model of Educational Experiences (Chapter One) the Teacher Characteristics construct and other constructs related to the system and school level contexts. Based on this conceptual model, the survey instruments developed for teachers assess a variety of contextual issues that have an impact on the nature and character of classroom instruction. As reported in Chapter ONe, much of the discussion over the classroom observations centered around those aspects that violated assumptions or were enigmatic. These led to the particular set of constructs described in this section. Recognizing the necessity for this kind of contextual information also led to development of the case studies which follow in the second part of this book.

SUBJECT MATTER ORIENTATION AND PEDAGOGICAL BELIEFS

Among the teacher characteristics that influence instruction are their beliefs about and orientation to subject matter and their beliefs about (subject matter specific) pedagogy. A teacher's particular perspective regarding a subject has an impact on how the teacher represents this subject and presents it to students. A teacher's ideas about how students best learn interact with subject matter resulting in different pedagogical approaches to lesson topics. (The concepts and their interaction are represented in the two models on pages 19 and 75.)

To assess some aspects of their ideas about subject matter, teachers are asked how important they think various practices or skills are in order for students to be good at mathematics (or science) in school. Six options are presented for mathematics such as "remember formulas and procedures", "be able to think creatively", and "be able to provide reasons to support their solutions" to which teachers choose "not important", "somewhat important", or "very important". Similar practices or skills are presented for science. A second item asks, "To what extent do you agree or disagree with each of the following statements?" A series of eight statements follow including, "Mathematics is primarily an abstract subject", "If students are having difficulty, an effective approach is to give them more practice by themselves during the class" and "Mathematics should be learned as sets of algorithms or rules that cover all possibilities". Teachers respond to these statements indicating whether they "strongly disagree", "disagree", "agree", or "strongly agree". Similar statements about science are asked of science teachers.

While teachers hold a variety of perspectives on subject matter, teaching and learning, these views do not interact in an abstract or theoretical realm but in response to specific situations and circumstances. This is an essential reason for the integrated nature of the CPF notion. Capturing CPF-relevant data involves having teachers reflect after the fact about typical situations normally handled as "flow" experiences without conscious analysis and deliberate reflection. A teacher's ideas about how to teach particular topics influences classroom pedagogy and would likely best be assessed by referring to a specific pedagogical situation or context. This approach was taken in the attempt to assess teacher's pedagogical beliefs in the Teacher Questionnaires. All pedagogical situations included were developed from the in-depth topics as part of the effort to obtain rich, complex, multidimensional integrated portraits for these selected topics. Figure 4-7 contains an example of one such item developed for the in-depth topic of proportionality for the Population Two Mathematics Teacher Questionnaire.

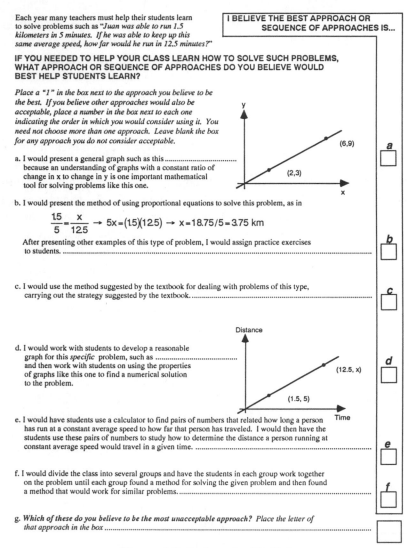

Each year many teachers must help their students learn to solve problems such as *"Juan was able to run 1.5 kilometers in 5 minutes. If he was able to keep up this same average speed, how far would he run in 12.5 minutes?"*

I BELIEVE THE BEST APPROACH OR SEQUENCE OF APPROACHES IS...

IF YOU NEEDED TO HELP YOUR CLASS LEARN HOW TO SOLVE SUCH PROBLEMS, WHAT APPROACH OR SEQUENCE OF APPROACHES DO YOU BELIEVE WOULD BEST HELP STUDENTS LEARN?

Place a "1" in the box next to the approach you believe to be the best. If you believe other approaches would also be acceptable, place a number in the box next to each one indicating the order in which you would consider using it. You need not choose more than one approach. Leave blank the box for any approach you do not consider acceptable.

a. I would present a general graph such as this ....
   because an understanding of graphs with a constant ratio of change in x to change in y is one important mathematical tool for solving problems like this one.

a

b. I would present the method of using proportional equations to solve this problem, as in

$$\frac{1.5}{5}=\frac{x}{12.5} \rightarrow 5x=(1.5)(12.5) \rightarrow x=18.75/5=3.75 \text{ km}$$

After presenting other examples of this type of problem, I would assign practice exercises to students. ....

b

c. I would use the method suggested by the textbook for dealing with problems of this type, carrying out the strategy suggested by the textbook. ....

c

d. I would work with students to develop a reasonable graph for this *specific* problem, such as ....
   and then work with students on using the properties of graphs like this one to find a numerical solution to the problem.

d

e. I would have students use a calculator to find pairs of numbers that related how long a person has run at a constant average speed to how far that person has traveled. I would then have the students use these pairs of numbers to study how to determine the distance a person running at constant average speed would travel in a given time. ....

e

f. I would divide the class into several groups and have the students in each group work together on the problem until each group found a method for solving the given problem and then found a method that would work for similar problems. ....

f

g. *Which of these do you believe to be the most unacceptable approach? Place the letter of that approach in the box* ....

Figure 4-7. Pedagogical Beliefs

DECISION MAKING

The sense of autonomy and perceived respect teachers feel can have an influence on teachers' perspectives and attitudes in the classroom. Several items were designed to assess these dimensions. One item, eliminated from the final version, asked teachers to rate the social status of teachers relative to other professions in their country. Another item asks teachers, "How much influence do you have on each of the following...". This is followed by four instructionally related decisions including "subject matter to be taught", "specific textbooks to be used", and "what supplies are purchased". Teachers indicate the amount of influence they have for each decision by checking "none", "little", "some", or "a lot".

In another item, teachers are asked to indicate what is their main source of written information when making four different decisions related to both curriculum and pedagogy. These decisions are "deciding what topics to teach", "deciding how to present a topic", "selecting problems and exercises for work in class and homework", and "selecting problems and applications for assessment and evaluation." Teachers indicate their preferred source for each decision from among a list of resources that includes the relevant national curriculum guide, the relevant regional curriculum guide, the school's curriculum guide, teacher edition of textbook, student edition of textbook, and an "other" resource.

PROFESSIONAL LIFE

Other items are designed to measure teachers' social and professional environment; the extent and demands of their professional responsibilities; and the extent to which teachers collaborate and support one another and are conversant with other resources in their instructional planning and preparation. One item asks teachers to indicate the number of individual school periods they are formally scheduled to teach each week. Following this is a list of various teaching and non-teaching responsibilities for which teachers indicate the amount of time they are formally scheduled each week. Another question asks teachers to indicate the approximate amount of time each week normally spent "on each of the following activities outside the formal school day". A list of eight activities follow such as "preparing or grading student tests or exams", "planning lessons by yourself", and "administrative tasks including staff meetings" for which teachers may choose "none", "less than 1 hour", "1-2 hours", "3-4 hours", or "more than 4 hours".

A teacher's professional life is also significantly affected by the amount of collegial interaction and cooperation. Teachers are asked, "About how often do you have meetings with other teachers in your subject area to discuss and plan curriculum or teaching approaches?" The options they may indicate range from "never" to "almost every day". An item from the school questionnaire also measures this aspect of teachers' professional environment by asking whether or not schools have official policies encouraging teacher collaboration and cooperation.

SCHOOL CONTEXT

The manner in which schools are organized to provide administrative support and oversight for teachers emerged as an important issue during the instrument development discussions as reported in Chapter One. Several items were developed for the School Questionnaire to address this issue. One lists several types of support staff and asks, "How many of the following are on the staff of your school?" Another lists 14 different tasks such as "establishing student grading policies", "placing students in classes" and "determining course content" and asks, "With regard to your school, who has primary responsibility for each of the following activities?" Respondents are asked to check a box after each activity to indicate whether it is the teachers' responsibility, the principal's, school or governing board's, or not a responsibility at all at the school level.

The school questionnaire also contains items designed to assess the degree to which teachers are able to specialize in teaching their particular subject matter, have time to prepare for their teaching, and have access to various instructional resources. Together with the portions of the Teacher Questionnaire crafted to assess aspects of classroom subject matter and pedagogy, these represent a concerted effort to capture through survey instruments the critical elements that come together to create the qualitatively different types of lessons observed.

LOOKING TO THE FUTURE

The TIMSS instruments developed by SMSO — the curriculum analysis, the School, Teacher, and Student Questionnaires — together with the student assessments represent an attempt to measure all the concepts included in the Educational Experiences model presented in Chapter One. At the time of this writing survey instrument development has been completed and the instruments already used in close to 45 countries as part of the main TIMSS survey. As the data from the surveys become available, opportunities will arise to explore more fully the issues around which the instruments were developed. Particularly for the six countries involved in the developmental project, it will be possible to examine, analyze and explore the data and the linkages among constructs in a richer way. The classroom observations previously completed as well as the case studies that follow in this volume can provide an invaluable context for comparative analyses among these countries. In two countries, Japan and the United States, additional information has been collected as part of the US TIMSS research effort. Videotapes of Population Two mathematics classes and extensive case studies of the education systems and settings in the two countries will provide invaluable additional contextual information for understanding and making sense of the survey data.

Were the survey instruments successful in capturing the qualitative differences found in the observation work? Were important aspects of instruction captured enabling the characterization of cross-national differences? Can the CPF hypothesis be corroborated on the basis of the data from the survey instruments? Soon these and other questions about relationships between curriculum and instruction can be explored cross-nationally.

There obviously are limitations inherent in the instruments developed for TIMSS — limits of survey methodology, elimination of items in seeking cross-national agreement on items, limits from not uncovering the best items or all important facets that needed measuring, etc. What was developed represents one approach to measuring specific concepts. Other approaches are both possible and needed. The approach presented here attempted to assess some aspects of classroom instruction obviously different across countries from observing classroom lessons. This approach has demonstrated promise, both through informal and formal pilots, for capturing and describing more carefully some of these differences. It is not yet known whether this particular approach will work in a large scale, survey assessment. One important aspect of this approach has been the attempt to conceptually derive all instruments on the basis of a model and to develop these instruments coherently so they could work together in creating a common portrait of classroom instruction.

One conclusion from the survey data may very well be that other methods such as observations, video studies, case studies, and more open-ended response items on questionnaires are essential in order to capture the important qualitative differences characterizing classroom instruction across countries. In developing the survey instruments, the qualitative data from classroom observations, and informal case studies available through the discussion of a multinational research team from the countries involved, proved to be invaluable. This discourse method serving the design and subsequent related analyses may only be possible when the discourse is centered around some sort of rich, qualitative artifact — such as observations, videos or case studies. Obviously such qualitative data may be essential to move beyond survey data in cross-national comparisons of what goes on in classrooms. If so, the SMSO team cannot stress enough the importance of the accompanying discussion and mutual examining of developmental data. It may very well be that multi-method assessment and investigation is essential for cross-national comparative studies that attempt to learn about schooling and the relationship of schooling experiences to learning and achievement.

Chapter 5:
Lessons from Lessons

In classrooms around the world, lessons are regularly built on curricular intentions and pedagogical strategies. These lessons unfold as teachers and students enact their well-defined roles in shared activities and move towards accepted goals and expected learning. The pattern of classroom interactions, activities, and the choices underlying them are familiar and commonplace within each country and culture. Classrooms represent characteristically patterned interactions among curriculum, teachers, and students. The heart of this story is the inextricable interaction of curriculum content dimensions with pedagogical approaches shaping instructional activity — interactions that yield qualitatively different lessons.

Earlier it was suggested that this was a story both simple and complex. The simple, unsurprising, and almost self-evident side of this story is that noticeable differences really do exist in classroom practices from one country to another. The far more complex side of the story comes in trying to uncover the sources of these very noticeable differences. The recognition of these characteristic pedagogical differences highlights the value of cross-national, cross-cultural classroom research.

Any country has diverse instructional practices within its classrooms, even when comparing classrooms for the same grade and subject matter. These differences are never inconsequential. Important consequences of this intra-country classroom variation often have been reported in the findings of educational research conducted within individual countries. Nevertheless, there are culturally distinct and nationally characteristic patterns in which curriculum and pedagogy intertwine within classrooms. These intra-country similarities are so great compared to inter-country differences that it is often possible to characterize typical national patterns in classroom lessons. These characteristic patterns, here labeled CPF, could only be brought to light through multi-national, multi-cultural investigation.

The true value of cross-national comparative studies is the opportunity to examine naturally occurring "experiments" — sets of educational phenomena varying far more across countries than they would within any one country. The ways in which different practices, approaches, orientations, and conceptions are combined in classrooms – ways in which topic, developmental, and cognitive complexity inextricably interact with content presentation and representation methods – reflect characteristic beliefs and practices in each country. If the truly important differences among characteristic national practices can be captured, they provide new perspectives on each nation's practices, approaches, and conceptions and provide an arena of variations which allow discovery of relationships among these things which would not be possible without a cross-national approach. One essential aspect of this hypothesis is that there are truly important differences to be uncovered. The SMSO work suggests that there are but that uncovering them is not easy. Quantitative and descriptive cross-national studies of classroom practice differences can be carried out by many methods including surveys. However, it is problematic whether such approaches identify important, fundamental differences rather than mere surface variations. Finding research methods to illuminate deep, essential and important differences is more difficult.

IMPLICATIONS FOR RESEARCH METHODOLOGY

Chapter One described the evolution of a research method for investigating classroom lessons, with their mixture of curriculum and pedagogy, in six countries. This was described as a "discourse" methodology – a process of a multi-national research team making sense out of data from a variety of sources and countries through a thorough exchange and interaction of the perspectives, insights, expertise, and surprising observations of each country representative. This approach was not only used to conceptualize what should be observed in classrooms but also to analyze the classroom observations conducted. However, the SMSO project was not created primarily to conduct classroom research but rather to develop survey instruments for TIMSS, a subsequent large-scale, cross-national investigation of mathematics and science teaching and learning. The classroom observations were conducted in developing and validating survey instruments for TIMSS. Two conclusions follow from the SMSO work. First, instrument development for cross-national studies requires grounding in conducting and analyzing classroom observations. Second, the method of multi-national discussion and exchange of ideas about a rich set of artifacts for common reflection — in this case, classroom observations — was a powerful approach to bring important, characteristic differences to light both for understanding the artifacts and for guiding instrument development.

Cross-national comparative research is extremely resource intensive, especially large-scale comparative studies. Perhaps because of this, the exploratory and developmental research that informs them has typically been drawn from existing research literature (that is, no fresh development was done); done within a single country; or done in rare and often brief international meetings with limited time for discussion, exploration and reflection. The result of this limitation is that issues from one national context are explored across a variety of other national contexts and cultures whether appropriate or not, or else the issues identified in brief shared discussion are those on which consensus can be easily reached, often because of assumed common understandings that would not stand up under deeper mutual reflection. In a very real sense such investigations begin with the assumption that the critical issues regarding the topic investigated are already known. The most blatant form of this is the "top down" delivery of research questions and designs with an invitation to countries to participate in these already formed strategies. At the other end of the spectrum is "bottom up", consensual identification of research questions and designs by the group of nations that have agreed to participate. Unfortunately, the result of this consensual approach, given the limits of participant discussion and the need for official representation for each country, is that studies are based on surface agreements whose depths have not truly been explored. For example, whether stemming from a "top down" or "bottom up" approach, studies might proceed assuming that the important varieties of instructional practices in mathematics and the sciences are already known. Research based on this assumption would seek primarily to look at how these differences vary across countries. As a result this research's focus would be on distributional differences, on matters that differ quantitatively. It would be assumed that these matters were the same as those that were fundamentally and qualitatively different. This assumption would typically remain unchallenged in the development of a large-scale cross-national comparative study.

What is assumed amounts to a series of tautologies: "education is education", "instructional practices are instructional practices", "homework is homework", etc., regardless of cultural and national setting. The discussion in Chapter One concerning "seatwork" and "homework" illustrated that such assumptions cannot be supported empirically. Chapters Two and Three provided other data challenging such assumptions. Survey research requires that its instruments be succinct and use precise, well-chosen words to ensure that important concepts are accurately and appropriately measured. The need for using few but well-chosen words demands reliance on shared meanings and understandings — or on the assumptions that meanings and understandings are shared and hence the survey instrument's concepts remain constant among participants. Obviously this is particularly problematic in cross-national research.

Translating survey research instruments is more than merely creating literal translations from one language to another. Literal translation does not ensure similar conceptual interpretations of survey instrument items. Questions may be interpreted differently by respondents and resulting data not be as comparable as assumed. To alleviate such difficulties, item wording often focuses on the highly specific and typical in order to not tax common conceptual interpretation by respondents and not to introduce hidden incommensurate responses by requiring much inference by respondents. The result is that such items may not accomplish the goal of capturing issues important to fundamental, qualitative differences. At times, those responsible for conducting such comparative cross-national studies have explicitly indicated that they would rather have less strategic but precise, commensurate data based on similarly understood items.

Is the link between more comparability and more response superficiality inevitable? Using traditional approaches to developing cross-national surveys the answer often unfortunately must be "yes". The developmental methodology used in SMSO indicates that these traditional approaches, and this seemingly inevitable link between comparability and simple responses, are not unavoidable but rather only difficult to avoid. The value of a discourse methodology — multi-national research teams, regular and prolonged mutual reflection and discussion, rich salient objects for common reflection, iterative refinement of common understandings — lies in the ability to have the assumptions brought to such investigations challenged and in developing shared meanings and understandings, meanings essential to creating carefully worded instruments that get precisely and comparably at fundamental issues rather than more surface features on which variation exists. In the context of a large-scale survey investigation, such a method ideally would be a preliminary complementary study on a smaller scale but with diversity representative of that typical for the larger study. It would be developmental and interact with more traditional, empirical pilot data collections and field trials.

The SMSO discourse approach confirmed that many important differences existed. Among the six countries, there were different types of teacher-student interactions; different ways in which materials were used; differences in what teachers did while students are working on activities alone or in groups; differences in how student groups were used during lessons; differences in how (or, whether) teachers lectured; differences in how teachers gave instructions; and differences in homework — differences in the regularity and frequency with which homework was assigned, in the nature and purpose of homework, and in the manner in which homework is integrated into sequences of lessons.

Only the shared SMSO dialogue brought to light how fundamental and significant these (and other) particular differences were. Only something like the shared SMSO dialogue could bring sufficient shared understanding of the important differences to allow precise and careful item wording that would allow comparable, low-inference responses providing data on these key differences. The discussion and shared reflection was necessary to make (common) sense out of the classroom observations but was also indispensable for developing the survey instruments. In conducting cross-national research, it is not possible to really know what questions to ask and how to ask them without significant cross-national dialogue of issues beyond translation; dialogue that establishes shared concepts and understandings to serve as a basis for careful item development that reflects those mutual understandings (and hence are likely to be comparably understood in the related larger scale study).

An assumption often made in past cross-national achievement studies is that curricula are the same or at least very similar (or, perhaps, irrelevant, or relevant only in the roughest terms to student achievement). Opportunity-to-learn (OTL) measures have often been included in such investigations but functioned primarily to allow for students' differential exposure to particular test items. Differences in curricular substance— particularly, in topic, developmental, and cognitive complexity — have often been ignored or considered infeasible to measure. Achievement differences among countries have been assumed attributable to factors other than the curriculum — for example, to differences in students' backgrounds or simple organizational features of schools and classrooms. Both Chapters Two and Three suggest strongly that subject matter content may be a far more important factor in explaining country differences in students' achievement than previously considered.

The important effect of subject matter on classroom practices and students' learning is not new. The observations conducted in the six countries emphasized the importance of subject matter content in understanding the nature and character of education and students' classrooms encounters with learning experiences. In comparative education research, these issues, and particularly what constitutes the subject matter being investigated, has not always been consistent across studies nor has it always been recognized as having a significant influence on student learning and performance.

SUMMARIZATIONS OF CPF

In Chapter Three it was suggested that curriculum and pedagogy — that is, subject matter and instructional practices — interact in characteristic and inextricable ways to yield qualitatively different lessons. The notion of CPF, characteristic pedagogical flow, was discussed as a conceptual key to characteristic lesson differences, at least among the six countries involved in the SMSO developmental project. CPF, as a general perspective, has implications for how lessons are structured and conducted. Teachers vary daily in how they structure and conduct lessons. Variation certainly exists among similar lessons within the same country. Even so, central to the CPF concept is the belief that, within countries, there are important commonalties in lesson structure and conduct, and that these obviously differ among countries. Generating characteristic country-level descriptions of classroom practice across both mathematics and science does not necessarily contravene the importance of subject matter in lessons. What is highlighted by country-level description is characteristic interaction between curriculum and pedagogy in lessons, presumably based in national beliefs and individual teachers' training and experience that leads them to share these beliefs. Interactions of curriculum and pedagogy based on such characteristic national ideologies and beliefs about education likely would be fairly consistent for both mathematics and science within a country, but vary among countries.

Characteristic pedagogical flow in lessons observed in each country can be described briefly based on the information presented in Chapters Two and Three. In France, lessons were characterized by formal and complex subject matter which the teacher actively organized and presented to students. The subject matter complexity stems from topic selection; emphasis on formal definitions, laws and principles; and the expectation that students would engage in theoretical reasoning and problem solving. Consistent with this, within each lesson segment —review, exposition or development, and application — discussion was based on observations and experiments as well as the discipline's theorems and principles.

Japanese lessons were characterized as built around a consideration of multiple approaches to carefully chosen practical examples or activities through which the teacher led students into an understanding of subject matter concepts and relationships. Subject matter complexity was primarily due to the way in which topics were developed and sequenced in the curriculum, and to the relatively focused and coherent development of them within lessons. Topic development in lessons generally occurred through subtly directed class discussions interspersed with periods for individual or small group reflection or practice.

In Norway, lessons were characterized by student activity, both individually and in small groups. Through these teacher prepared activities, students were expected to come to understand basic information and facts about subject matter. Cognitive complexity, what students were expected to do, was central to the uniqueness of these lessons emphasizing procedures and processes — performing operations in mathematics and using scientific procedures and investigation in the sciences. During lessons students worked independently or in a small group much of the time with teachers available as needed. Such independent practice or exploration times tended to be longer, fewer and less integrated into the teacher's explanation of the lesson than in either France or Japan. Consistent with the cognitive emphasis found in textbooks, much of the discussion in lessons involved explanations about the learning activities, and of basic facts and vocabulary.

Lessons from Spain were characterized by the teacher introducing students to formal and complex subject matter. Similar to France, the complexity of the subject matter stemmed from the topics selected; the emphasis on formal definitions, laws and principles; and the theoretical reasoning and problem solving expected of students. Unlike France, textbooks played an important role for both teachers and students in lesson topic development. Review of previously covered topics, especially as related to students' homework, occurred at a relatively conceptual level and played a significant role in lessons' conceptual development and overall structure.

The Swiss lessons were characterized by student subject matter exploration and investigation through learning activities and teacher demonstrations. Similar to Norway, there was an emphasis on students' responsibility for their own learning. They were expected to come to understand subject matter information and facts through the learning activities prepared by the teacher. There was comparatively greater diversity in what students were expected to do (cognitive complexity) than in some of the other countries. Textbooks, however, showed a preponderant emphasis on students' knowing, using and understanding information. Consistent with this emphasis, teachers often asked students for their observations, conjectures, or conclusions during lesson development.

Lessons from the US were characterized by teachers presenting information and directing student activities and exercises. The multiplicity and diversity of both topics and activities was a unique feature of US lessons. Both teacher and student activity tended to emphasize the basic definitions, procedures and concepts of subject matter. Consistent with the cognitive emphasis found in textbooks, the preponderance of lesson discussion involved information about procedures, exercises and basic facts. Content complexity in lessons stemmed from an emphasis on knowing technical vocabulary and definitions. Periods of independent student practice or application were common and appeared to function in a manner similar to that found in the Norwegian lessons. Such periods contributed to the low content visibility in some observed lessons. These general characterizations seem to have held despite greater variability and diversity among the US lessons than among those from any other country.

LOOKING TO THE FUTURE

Instruction differs qualitatively among countries. The implications of this simple fact go far beyond the commonplace notion that countries emphasize different variables in instruction—the idea that students do less homework in Japan, for instance, or that they are evaluated through rigorous, external examinations in France. The presence of qualitative differences in mathematics and science instruction across the six countries points in a different direction, suggesting that more complex factors than typically assumed may be at work in accounting for instructional differences.

Mathematics and science, unlike culturally embedded subjects such as history and language, are often thought to be a-cultural. For example, many believe "numeration is numeration" — the concept is the same across all contexts. The same is believed true of science. It is considered not to matter whether the notion of molecule is encountered in Japan or in France, in the fourth grade or the eighth — in all cases many consider it to carry the same meaning, a meaning that can be unambiguously measured in attainment tests. One can argue that if there is something universal about mathematics and science content, there should be something universal about the way this content is presented to students. Our results, of course, suggest that this second assumption needs re-evaluating. Logic suggests the first does also.

Countries have developed their own ways of engaging students in the substance of mathematics and science. There appears to be strong cultural components, even national ideologies, in the teaching of these subjects. The French stress on formal knowledge seems less surprising after considering how they have organized their educational system — in France, upper secondary school teachers identify strongly with their disciplinary counterparts at universities. There is a sharp demarcation between teachers at this level and those who work with younger children. This sends a strong signal about the importance of formal disciplinary knowledge throughout the pre-university system, a signal likely to affect how teachers approach school subjects at the primary and lower secondary level. In Japan, great care is taken at the primary level to expose children to important disciplinary concepts in an informal but carefully orchestrated way. This appears to reflect a dual focus at the primary level on ensuring that instruction meets the needs of the child but, simultaneously, prepares the child for more rigorous and formal study at the lower secondary and upper secondary school level. US educators, in contrast, appear bent on prolonging childhood as long as possible, at least as evidenced by the tendency to extend exposure to more basic and early-introduced topics in mathematics and science well into lower secondary school.

If, as the SMSO classroom data suggest, there is a strong cultural component to mathematics and science teaching, it may be time to rethink and refine the "world as a laboratory" metaphor that has colored much thinking about cross–national educational research. Thorsten Husén, a founder of IEA, appears to have originated the "laboratory" metaphor as a way to encourage cross–national comparisons in reading, mathematics, and science. If one thinks of the whole world as an educational laboratory, he argued, it makes sense for researchers to distinguish between those "experiments" that have proven more and less successful in elevating achievement. Once this is done, the factors involved in the more successful experiments can be identified and studied. The results of this analysis should prove useful for those interested in replicating the results of the experiment (i.e., in attaining the same high level of achievement).

Viewing mathematics and science teaching through a cultural lens raises questions about the relevance of the laboratory metaphor. A laboratory, educational or otherwise, has certain attributes that make it extremely useful from a scientific standpoint. For example, it is a carefully controlled environment, which is important when one attempts to test a well–defined model or theory. Without the ability to isolate effects by manipulating variables systematically and individually, it is impossible to engage in causal analysis. One cannot rule out the potentially contaminating effects of seemingly extraneous or irrelevant variables.

Generalizing the laboratory model, frequently researchers involved in cross–national achievement research have assumed that it is possible to isolate the variables that contribute to students' educational achievement. Variables may differ quantitatively from one locale to another but the essential nature of the variable remains unchanged. According to this assumption teacher subject matter knowledge, as one example, can be operationalized in the same way around the world. The same is true of seatwork assignments or teacher feedback. Is this a realistic assumption? Current research suggests it is not. Anthropological research in the US demonstrates that teacher feedback, if it is to be judged effective, must be geared to the cultural norms of the recipients of the feedback. Native American students respond much less favorably to individualized feedback, preferring to view themselves as members of a group rather than as a collection of competing individuals. Individualized feedback, on the other hand, is well suited to middle class members of the mainstream culture. Correlating teacher individualized feedback and student achievement without taking into account this important cultural difference would be unwise in light of this research. Similar culturally sensitive interpretational and contextual issues underlie cross-national achievement comparisons.

The laboratory metaphor is too comfortable. It suggests a degree of closure about the nature of the educational enterprise that misrepresents our current level of knowledge. Arguing for a totally individualized approach to educational achievement seems equally inappropriate. Acceptance of this more relativistic conception of curriculum and schooling creates its own set of difficulties. If educational practice can only be evaluated relative to a country or a culture's norms, then cross-national comparisons seem impossible and unfruitful. There would be no metric – even at a deeper conceptual level – that provides a common framework for analysis across systems. One country, in this case, would seem to have little chance of learning from another.

This relativistic position is too extreme just as the laboratory position is too mechanistic. We must recognize empirical differences but also empirical similarities. The lesson of the SMSO analyses is not only are there significant qualitative differences among characteristic pedagogies observed in lessons, but also a variety of structural and functional homologies. For example, in all SMSO countries a common function of teachers was to provide feedback on student performance. In fact, the structural and functional similarities in mathematics and science instruction were enough that, for example, a Japanese observer could enter a classroom in any of the other countries and without knowing the language (or even the country she was in) be able to determine whether a mathematics or science lesson was in progress. Given such fundamental similarities as well as striking differences, a middle ground solution seems possible. A nation's educational system is neither a laboratory in the sense of a sterile, manipulable environment in which a well–specified model is tested nor, we believe, is it a unique structure that can only be understood in its own terms.

The research presented here suggests that there is a middle ground. This middle ground is based on the intuitively reasonable assumption that conceptually robust variables, such as effective teacher feedback, can be operationalized in different ways but retain fundamental identities in functionality, thus allowing for meaningful comparisons across cultures. Thus, commenting on individual student performance may be one of several ways that teachers could provide effective feedback to students. Group feedback, in cultures where this is considered more appropriate, serves a conceptually similar purpose inasmuch as it signals a teacher's interest in or concern about student performance. Understanding these differences better would still allow teacher's use of feedback to be a significant part of explaining achievement results, but not through simplisticly gathered or interpreted data.

Managing the instructional process is a second example of a complex notion defined in different ways while serving the same overall purpose — structuring the instructional process in a systemic context. For example, in France, regulation is achieved by having low achieving students repeat classes. In the US, regulation takes a different form and is typically managed through a tracking system that tries to match content difficulty and student ability.

Methodologically, cross–national research is often designed with a general model of educational achievement in mind. This model may then be modified based upon pilot work or information provided by local sources of information. As a result of this developmental work, some variables are dropped and others retained. In fact, this was the initial SMSO approach. We attempted to adapt our common instruments to unique, country–by–country circumstances, an improvement over more traditional "one size fits all" approaches but still inadequate. The general model presented in Chapter One (Figure 1-2, page 19) worked fairly well across the six countries involved in the study — with one notable exception. It did not capture very well the characteristic pedagogical flow described by our Swiss representative and the cases presented by the Swiss researchers. Models such as the one we used, of necessity put variables in "boxes", suggesting a degree of boundedness, self–containment or dis-integration that simplifies reality to the point of sometimes misrepresenting it.. In the Swiss example, there was some uncertainty about how even to define a unit as basic as a school. The bounds of the school in that country are much more permeable than in most other countries — there is no headmaster in primary schools in Switzerland, and parents are much more intimately involved in their youngsters' schooling compared to countries such as France, Spain, or the United States.

Results of the present study suggest that cross–national researchers may find it more productive to employ a "specific to general" as opposed to "general to specific" model of instrument development. That is, it may be more fruitful to begin the difficult work of instrument development by starting with individual cases of teaching and learning, analyzed on a country by country basis in a way similar to what was attempted here. Cases serve as a common text in the process of negotiating shared understandings of what is happening educationally across systems.

Researchers from different countries bring different lenses to bear in viewing and interpreting this common text, thus bringing to the surface points of agreement and disagreement. Further discussion helps clarify the extent to which the agreements and disagreement represent varying interpretations of the same underlying phenomena or a deeper, more fundamental variance in conceptual viewpoint. In the case of varying interpretations of the same phenomena, it is possible to arrive at multiple definitions of what is, at root, a common construct. We propose an approach to instrument development that attempts to identify structural or functional homologies across countries (in educational phenomena of interest). These would serve to identify constructs. The challenge would be to recognize that measuring such constructs based on homologies cannot be done by strictly following traditional methodologies. The notion of "standardized measures" must be re-examined for context-sensitive educational evaluation. This allows for the development of more valid cross–national instruments which, in turn, are necessary for any valid comparison across national educational boundaries.

Instruments developed in the future, if in fact they are successful in capturing the sort of underlying phenomena discussed above, may look very different from those we currently use. They may marry more traditional, quantitatively-oriented items arraying classrooms or teachers along a single continuum—thus representing differences in simple comparative terms—with sets of items which, considered as a whole, portray group differences in qualitative rather than in quantitative terms. The first approach assumes that differences between groups are unidimensional while the second approach views differences more complexly, as patterns of responses represented in a multidimensional space. The latter approach is amenable to sophisticated statistical analysis, thus enjoying the advantages typically associated with quantitative methodologies.

A good example of a construct that can be used to capture qualitative differences in instruction is that highlighted in this book — characteristic pedagogical flow. This complex construct attempts to capture patterns of interaction between content and pedagogy at as broad a level as possible. It has proven useful as a heuristic in generating survey items we believe will allow the capture of diverse instructional patterns or typologies. The patterns or typologies identified in this way can then be related to more traditional quantitative measures such as mathematics and science achievement.

Attempting to understand educational systems as they exist in the real world is a difficult enterprise. Negotiating a common understanding across national boundaries requires investments of time, energy, and good will on the part of all involved. Such investments should lead to better instruments which, in turn, should yield more reliable and informative survey data. The latter is essential if we are to develop a deeper understanding of how other nations organize their educational systems — and what that means for student achievement. It is by looking through the lens of other cultures that we come to understand the nature of our own strengths and weaknesses. A realistic view of the latter is a necessary prerequisite for informed policy making.

Given the complexity of characterizing curriculum and instruction discussed here, it may seem to some better simply to avoid these more problematic aspects of education and focus on seemingly simpler matters such as measuring student attainments, which may well be comparatively insensitive to curricular and instructional factors anyway. Such a belief certainly misunderstands the difficulty of understanding student attainments in context, assuming a more simplistic and de-contextualized approach than is warranted. Most importantly, however, curriculum, schooling and instruction are not "messy" parts of education to be avoided. Of what use is it to know comparative attainments of nations' students without understanding how this is related to their schooling? Schooling is the issue, the focus, of education. It is schooling that leads to student attainment. Focusing solely upon cross-national differences in student attainments presupposes that the entire schooling process – educational goals, curriculum, and pedagogy – are invariant across countries. Such a focus precludes the possibility of constructing useful explanations for these differences – explanations necessary to inform policy and practice. Student attainment is but one of the indicators by which comparative effectiveness in schooling can be assessed. In cross-national comparative studies meant to be relevant to nations' efforts to improve their children's educations, one must keep firmly in mind which is the "tail" and which is the "dog". Schooling shapes attainments; only when used as part of a cyclic improvement process should attainments help shape schooling. These simple facts must be remembered in the design and execution of any serious cross-national studies of education.

PART II
Case Studies

Japan

France

Norway

Spain

Switzerland

United States

FRANCE:
CASE STUDY

FOREWORD

The French educational system consists of three cycles: primary school, a 5-year cycle which pupils can enter as soon as they are 6; lower secondary school (4 years); and upper secondary school which offers general, technical or vocational courses. Quite a lot of children attend pre-primary school, beginning at age 3, where they are prepared for primary school. Nearly all 5-year olds go to a pre-primary school although compulsory education does not begin until age 6.

At the end of primary school, all pupils enter lower secondary school known as collège unique, where all follow a common curriculum. The collège is, therefore, a comprehensive school. At the end of collège, students are sent either to work (if they have reached the age of 16 and their attainments are very poor) or to a general, technical or vocational (in decreasing order of prestige) secondary school. School attendance is compulsory until age 16 which corresponds to, theoretically, the first year of secondary school. At the end of collège (Grade 9) there is a non-compulsory exam (the Brevet). This exam is not required for continued education but is a common requirement for employment if one does not obtain another exam certification. At the end of upper secondary school (lycée), students take the Baccalaureat, technical or vocational exams.

There are no exams required for passage into the next grade, or cycle. This decision is made by all teachers of the class (or by the teacher in primary school) together with the headmaster. The decision is based on student's achievement in all subjects as assessed by specially designed exercises, mostly written, which have been given throughout the school-year during a class. Although the decision is collective in secondary schools, French and mathematics teachers have particular influence. When a student's attainments are considered insufficient, he is held back. Repeating a year is quite common, especially in first-cycle secondary schools, in spite of efforts aimed at reducing it. Although parents do have some influence on the retention decision, it is especially common at the end of grades 8 and 9.

Teachers at the primary school level receive general academic training. Until recently, the minimum required was the baccalaureate plus two additional years of university study. Since 1992, the requirement is the baccalaureate plus four additional years of university study (that is 3 years at the university and one devoted to theoretical preparation for the final exam). Both requirements are followed by professional training. Teachers may then teach all subjects except in special areas: foreign languages, music, art, or physical education.

At the secondary level (collège and secondary school or lycée) teachers are given advanced training in one subject. The training is supplied by the universities in their disciplinary departments (mathematics, physical science, etc.) rather than in departments of science education. Thus, teachers are first of all specialists in their subject. At the end of university studies they take a competitive exam in their subject and, if they pass, receive pedagogical training for one year, which includes time in the classroom. At the end of that year, they take another competitive exam in the presence of a jury.

SCIENCE, POPULATION 2

The following describes a science fourth year class in a lower secondary school (eighth grade). The class is made up of 30 students with a range of backgrounds and academic proficiencies with as many boys as girls. Such mixed classes are the norm in French education. The teacher, like all physics teachers, teaches chemistry as well as physics.

A typical science class would include all or some of the following sequences: detailed, collective correction of homework; presentation of the subject of the lesson, starting from observations, documents, personal experience, etc.; experiments (demonstrations) performed in front of the class; a search for laws and relationships between observed phenomena; conclusions; application exercises and homework.

The class under observation lasts 90 minutes with a 15 minute pause. The topic under discussion is electricity. The flow of the lesson may be divided into several sequences.

Science and mathematics have a national curriculum which is set by General Inspectors or the National Curriculum Board (Commission Nationale des Programmes) who append some recommendations. The curriculum is mandatory for all teachers. Official instructions also specify the number of periods that are to be allotted each subject (e.g., natural science or mathematics) for each grade.

At the lower secondary school level mathematics and the natural sciences (i.e., biology and geology) are all taught from grades 6 through 9 (i.e., throughout the cycle). The same teacher teaches both natural sciences and chooses the sequence and emphasis given each science over the course of a year. In a similar manner, the physical sciences (i.e., physics and chemistry) are introduced the third year and are taught in addition to the natural sciences for the last two years (grades 8 and 9)(see Monchablon, 1995).

SEQUENCE 1: 15 MINUTES

The teacher corrects an exercise given at the end of the previous class, which the students are expected to have solved at home. Several of them volunteer their answers, one of them comes to the blackboard, and writes his answer with the help of the others when necessary. The teacher intervenes to ask for clarification for certain words or justifications. Students answer spontaneously; the teacher does not necessarily select a particular student to give an answer.

Homework, in the form of written exercises, is the rule between classes. The exercises do not systematically consist of the reproduction of exercises done in the classroom. They often call for more reflection, generally taking one hour to complete. The work is mandatory. Teachers walk the rows in order to check that work has been done and students who repeatedly fail to do their homework are often penalized.

Teachers are free to organize progress, to choose their own methods of teaching and assessment, without any interference from colleagues. As there are no Department Heads in lower or upper secondary school, two math teachers operating in the same grade in the same school are likely not to progress at the same pace or to use the same assessments methods or tools.

The teacher is sole master in the classroom, even at the primary school level. The headmaster, let alone another teacher, does not have the prerogative to interfere with the teacher's pedagogy. Only an Inspector can do so. Such visits and interventions by Inspectors are common in primary schools. In other schools, teachers are visited by an inspector only once every three years for one or two hours and often, even less than this. Even though pedagogical team work is recommended, no teacher can be required to participate.

Teachers make a point of creating their own teaching outline (of course with the help of textbooks). Therefore, where there are teachers' guide-books, they contain little else than the right answers to exercises, and have but little influence on teaching practices. Students use their textbooks, most of the time, for illustration and exercises.

SEQUENCE 2: 10 MINUTES

The teacher introduces a new exercise related to a concrete situation. Again there is questioning between the teacher and students about the correct answer to the problem. Students participate by raising their hands.

The objective, at this stage, is to check students' mastery of concepts which will be used during the lesson; in this case, alternating power, voltage and frequency. There is frequent intervention as the teacher refers to knowledge that is supposed to have been acquired and is a prerequisite to students' understanding of what follows.

The teacher continues asking questions about the previous lesson. Discussion is voluntary and the students can express themselves freely.

SEQUENCE 3: 10 MINUTES

The teacher announces the subject of the lesson, and writes this on the blackboard:

The conveyance and distribution of energy

1. The Transformer

1.1 Study of document

The blackboard is essential in teaching. The teacher is expected to put up the lesson outline. Demonstrations must also be written out on the blackboard as well as any new terms and experimental drawings corresponding to the equipment on the teacher's desk. Students must take notes and the blackboard is a great help. The outline provides the necessary structure for the lesson.

The teacher asks the students to open their books to a page where they will find a document showing how electricity is conveyed to the customer, starting from a generating plant and, after some transformations, through successive reductions of voltage.

There are no official textbooks. Publishing companies develop and publish textbooks and propose them to schools and teachers. Most of the time, these textbooks are written by teachers under the authority of a general inspector. Textbooks are selected by a team of teachers for each subject and, in primary and lower secondary schools, are purchased by the school for students' use.

The teacher asks the students to describe the progress from the plant to the users and asks what the different voltages are, which is the highest and which the lowest. Several students take part in the ensuing discussion incorporating their common knowledge of transformers. The teacher then explains the notion of energy lost in the process and the necessity of transformers. She thus comes round to the lesson itself.

Though the lesson is effectively managed by the teacher, exchanges between students are frequent and encouraged. The teacher intervenes to rephrase questions and answers if some students appear not to have heard or understood.

SEQUENCE 4: 20 MINUTES

The teacher then proceeds to the lesson proper and writes on the board:

1.2 Description of the transformer

1.3 Measurements

She then asks a series of questions which the students answer. The word "transformer" has been used spontaneously by several of them in the course of observing the illustration in the book.

The teacher proceeds to take apart a transformer and shows the different parts, particularly the soft iron core. The transformer is a large one that can easily be taken apart and has been especially devised for teaching purposes. All the students in the class can easily see it from their places.

Whenever students conduct practical exercises or investigations in the sciences (whether physics or natural science) these are held in specially equipped classrooms where very strict safety norms must be observed and where practical work is possible. Students' tables are tiled and have gas, water, and electricity available (the latter is often supplied by appliances placed on the wall next to the table). Most of the time, lessons are held in rooms where such special equipment is available at one large table which only the teacher uses.

The teacher can always use a large blackboard table supplied with gas, water and electricity which enables her to perform experiments in front of the students. She uses over-sized apparatuses whenever possible (made especially for schools) so all the students can see what is taking place at the desk. She can also use overhead projectors, cameras, or VCRs. Schools always receive a certain sum of money for such equipment, so, in theory, there are no differences between schools as regards the possibility of using experiments for teaching purposes. In lower secondary school (contrary to upper secondary), laboratory staff is very rare and teachers have to prepare their own experiments and instruments for all the students before the class begins.

The teacher launches an experiment consisting of measuring voltage with the transformer and voltmeters. A diagram of the experimental apparatus has already been drawn on the blackboard. The lesson follows as three students participate in the experiment by taking measurements. Their values are placed on the board. The teacher then introduces the symbol for transformer on the board and defines the terms "primary" and "secondary". The sequence ends with the teacher dictating a final summary of the demonstration and the conclusion which is given by the students.

Though the students are used to taking notes, teachers at the lower secondary school level make sure that students write what is essential by dictating a summary.

SEQUENCE 5: 25 MINUTES

After the 15 minute break, work begins again with a discussion about the results of the measurements.

The final part of the lesson begins when the teacher writes on the blackboard:

1.4 Transformation rate

She defines what a transformation rate is and assigns an exercise in which the number of coils is specified, together with the primary and secondary voltages. Students try to determine the link between the number of coils and the corresponding voltage.

The teacher gives the final conclusion, namely the relation existing between those two rates. She then announces that the lesson is to be completed next week with the study of the conveyance of electricity.

She assigns homework for the next week: learning the lesson and the summary in the book. Students are asked to do exercises from the book, as well as study the illustration observed during the class. She also announces that a test will take place the following week.

The teaching of physics relies heavily upon experimentation and measurements whenever materially possible. Equipment specially designed for schools frequently permits this. Students are trained to conduct and to discuss these experiments amongst themselves. They can also perform such manipulations during special sessions, called "practical work". The aim of a physics lesson is to deduce laws, and relationships between observed phenomena, using a minimum number of new concepts. The goal is to have students understand these laws and principles through observations of concrete examples. Students are not only required to describe what they observe. They must also search for the explanation and laws governing the observed phenomena which is first described in everyday language. Then in increasingly specific terms, students arrive at the scientific rule or principle. Moreover, application exercises are very frequent with a view towards reinforcing learning.

MATHEMATICS, POPULATION 2

The observed class is a grade 8 lower secondary school mathematics class. The lesson takes place in an ordinary classroom. The students sit in pairs at tables with chairs. The tables are large enough for them to write and draw with ease. They take notes throughout the lesson.

Math is taught in ordinary classrooms: tables and chairs for the students for easy note-taking and figure drawing; a desk and large blackboard (several meters in width) for the teacher. Most of the time, math takes place in the room where the form (i.e., group of students) has most of their classes. The teacher may choose the arrangement of the tables (in a U, in rows, in a circle, etc.). The number of students in a class should not exceed 25 in lower secondary school but their numbers are actually higher in many cases.

The subject of the lesson is operations on powers and has previously been introduced in former lessons. The teacher does not use the students' textbook except as a source of further exercises and organizes the lessons himself. The lesson begins at 11:00 and lasts 55 minutes and is divided into three sequences: correction of homework, lesson presentation, and application exercises.

SEQUENCE 1: CORRECTION OF HOMEWORK (15 MINUTES)

The teacher announces that she is going to correct the 4 exercises given at the end of the previous class. A student is sent to the blackboard to give the correction of the first exercise from the homework. During the correction of the homework problems the student is expected to justify reasoning and calculations.

This sequence devoted to the correction of exercises done at home is quite typical of what students are required to do in several ways: first, the number of exercises given (e.g., 4); and second, the degree of difficulty. The teacher reminds his students that the rules presented and elaborated upon during the previous class have been applied to simpler examples. Homework is not merely a simple repetition of work done in the classroom: it expands upon and develops the latter. This is the reason why detailed correction of this work is the rule.

SEQUENCE 2: PRESENTATION OF THE LESSON (15 MINUTES)

The chapter on powers was introduced in the previous class. The teacher reminds his students that he has already dealt with the fundamental definitions and formulas.

He writes the beginning of his outline on the blackboard:

III. Complements

1. Exponent zero

The meaning of a number written with a negative exponent is written on the board and copied by the students into their notebooks. The students use their calculators to verify that any number raised to the 0 power is equal to 1. The teacher walks among his students to see that calculators are properly used. The teacher then has the students verify the conventional notation by applying what they know about division of numbers with exponents to the special case of division of two numbers with the same exponent.

The teacher continues writing his outline on the blackboard:

2. Negative exponent

The teacher then uses the same strategy as used in developing the idea of a zero exponent. After demonstrating the meaning of the conventional notation for negative exponents on the blackboard, he asks a student to justify the conventional notation by applying what the students already know about division of numbers with exponents. The student works at the board with the assistance of other students. This process is carried out at the board several times. Using this type of routine, the teacher regularly has his students verify the meaning of common notations and formulas.

During the course of this sequence, new content and new rules are proposed to students. What makes it such a typical sequence, is the use of calculators together with a demonstration or proof. He not only uses the rule (formula) but also wants the students to understand the process. In tests, teachers will not allow students simply to give results without explicitly applying a rule.

SEQUENCE 3: APPLICATION EXERCISES (25 MINUTES)

The teacher assigns two exercises from the students' book. He insists that they read the exercises very carefully. This is something that teachers say very often to their students.

The students do individual work for 15 minutes. The teacher walks between the rows and checks each student's work individually. He points out incorrect results and reminds the students of the 5 rules they must know.

The teacher corrects operations on the blackboard, showing that several methods can be used. He concludes the lesson by assigning homework for the next class: 2 collections of exercises from the book each comprised of 5 items.

Contrary to observations carried out during the physics class, student-to-student exchange is infrequent and teacher-to-student interaction is initiated by the teacher who does most of the talking. Students interject comments during correction or ask spontaneous questions when the teacher looks over their shoulder to see what they are doing during sequence 3. But the teacher is also sensitive to students' uncertainty when confronted with a complex task which leads him to emphasize and explain the mistakes they are likely to make.

This lesson is marked by important individual work by the students. Each one of them must solve the exercises alone even if interaction with neighboring students is allowed. Individual work is always valued in France. After a period of approximately 5 minutes, the bell rings. As students gather their belongings to move to their next class, the teacher reminds them of the homework they need to do.

JAPAN:
CASE STUDY

FOREWORD

Japan's Ministry of Education revised the national course of study in 1988. The national objective of elementary mathematics is described as follows:

> *Students will acquire fundamental knowledge and skills in numbers, quantities, and geometric figures. Based on this knowledge, students will develop an ability to estimate and to think logically about daily events around them. Students will also develop an appreciation for the mathematical processes involved and be able to utilize these mathematical processes in daily life.*

The content of elementary mathematics is divided into the following four categories:

> *A. Numbers and computation*
> *B. Quantities and measurements*
> *C. Geometric figures*
> *D. Numerical relationships.*

In the new course of study in elementary mathematics, two key points of revision are "enriching a sense for numbers, quantities and geometrical figures" and "estimating appropriately about numbers, quantities and geometrical figures with respect to specific content areas". Key points of revision relevant to teaching strategies include "enriching problem solving skills based on previous learning activities", "using activities for concrete operations and thought experiments" and "educating for individuality". Students' use of calculators and computers is also recommended. Generally, students begin to use calculators in the fifth grade.

There is also a framework to record students' progress recommended by the Ministry of Education. This framework is used to make a semester report for parents. The framework tracks the progress of the following areas of learning:

> *a) Interest, motives and attitudes*
> *b) Thinking*
> *c) Knowledge and understanding*
> *d) Skills and procedures*

This view of learning ability represents a new focus on students' "interest, motives and attitudes" as a core driving force of learning. Previously knowledge and understanding was considered the number one priority.

MATHEMATICS, POPULATION 1

The lesson described takes place in a fourth grade classroom (age 10) in a regular primary school. There are 18 boys and 16 girls in the class. The topic of the lesson is problem solving in mathematics.

Kumi, an experienced elementary teacher, planned to have three hours on problem solving activities. This lesson is the first of the three in this unit. She wants the students to appreciate the idea and use of mathematical functions in problem solving.

Kumi began her preparations for this class the day before the lesson by making several pieces of equilateral triangles out of cardboard. She will use these as visual aids in her presentation and discussion with the class.

All textbooks are approved by the Ministry of Education. Textbook companies publish a teacher's manual that provides detailed teaching plans and exercises for students. In mathematics, teachers often ask students to purchase one or two drill and practice workbooks which are used in class and at home. Main instructional resources used in classrooms are textbooks, supplementary materials, hand-outs made by teachers, manipulatives, etc. The use of manipulatives is commonly observed in elementary mathematics classes.

The time is 10:45 and students are coming back from the school yard after recess. It takes a few minutes for everyone to sit down. They sit in rows of two, girls sitting next to boys. After all of the students are seated, she makes contact with one student to start the lesson.

The student says "Stand up" (Kiritsu) and "Bow" (Rei).

All students bow to her and she bows to them.

The student says "We start the math lesson" and all students sit down.

Kumi then starts to talk to her students.

The size of classrooms is legally defined. Therefore, many classrooms in public schools look very similar. There are two doors, one in front and one in back on the same side of the classroom. There are windows on the opposite side. The blackboard is located in the front of the classroom and a small blackboard is located on the other side. The teacher has her desk in the front of the classroom. Usually, elementary classrooms are decorated with students' drawings, calligraphies and "han" exhibits (described later).

There are usually about thirty students in elementary classrooms. It is reported that the national average number of students per class was 30.8 for public elementary schools, in May 1988. If the number exceeds forty, the class is divided into two classes.

Kumi begins the lesson by putting a 5 cm equilateral cardboard triangle on the board and asks the class to estimate the perimeter of the triangle. After receiving a correct answer from one student (15 cm), she places a second triangle on the board next to the first one. The task now is to find out the perimeter of the two adjoining triangles (20 cm).

Kumi then asks her students to think about a problem: what would the perimeter be if we placed 10 equilateral triangles adjacent to each other? She writes this problem on the board and asks students to copy the question into their notebooks. She waits a few minutes as students work individually on this problem. She notices that some of her students are struggling with the task.

The next part of the lesson is an interchange between Kumi and her students as they discuss the solution to the problem. Students raise their hands when they think they have the correct answer to a question. At this point in the lesson, all answers are accepted.

Students are then asked to form their "han" (group) to discuss the solution to this problem. They are told to think about strategies they might use to solve the problem.

Students are organized into groups of four to six called a han. The han moves their desks together when working in groups. During the lesson, they will often work together, cooperate and discuss. This group also functions as one unit of school life: they eat lunch together, clean up the school together and share classroom responsibilities together.

After a period of a few minutes she gives her students a few suggestions to try and asks them to work individually, writing their work into their workbooks. About 10 minutes are spent in this activity. If students are having problems, Kumi gives them small triangles to work with. She suggests that students make a table and if they have solved the problem, to work on finding an alternative solution.

At the end of this task, Kumi leads a group discussion on possible solutions to the problem. Students are asked to write their solutions on the board. A number of different solutions are presented, some correct and some incorrect.

The discussion continues as students are asked to concentrate on similarities between those solutions which are correct and eventually to explain these solutions. Kumi is trying to help her students understand that there are multiple solutions to the same problem and that there are similarities in the correct solutions. In this way she is approaching the idea of functions. Students are asked to summarize the discussion in their notebooks.

As with other Japanese teachers, Kumi is trained to use the following sequence for making lesson plans, although variation does exist:
    Introduction of content (Dounyu)
    Development of content (Tenkai)
    Summarization of lesson (Matome)
Primary teachers teach all of the subjects in their class. However, some subjects such as music, arts and crafts, and physical education may have special teachers. Teachers in the same grade may also switch hours with each other. For example, a male teacher may teach physical education for two classes and a female teacher may teach music for two classes.

In October, 1986, it was reported by the Ministry of Education that elementary teachers taught 22 hours per week. In addition to these teaching hours, elementary teachers are involved in extra curricular activities and other school events. This means that they are very busy during the week and it seems hard to have time to prepare lessons. However, lesson plans and other instructional resources are available both from curriculum guides, published by the Ministry, and textbook publishers to assist teachers in their lesson planning.

Kumi concludes the lesson by summarizing what they have done during the lesson. She says, "We have solved a problem involving the perimeter of triangles and have found that there is more than one solution to the problem. Each solution has its own merit." She then tells them about the next lesson they will have.

The lesson ends as the students and teacher bow to each other.

Kumi collected the student's notebooks and checked each student's work after the lesson. Some students did not seem able to follow the day's lesson, but many students seemed to understand what went on. Today's lesson went very well, but sometimes it is very difficult for her to encourage students to think. Kumi often confronts two challenges: one is the problem of involving all of the students in their concept development; the other is the problem of covering all of the topics in the crowded mathematics curriculum. This seems to be a constant struggle for teachers in Japan.

SCIENCE, POPULATION 2

The lesson described takes place in a school located in the center of downtown Tokyo. All students wear school uniforms at the junior high school level in Japan. In elementary schools, everyday clothes are worn. The atmosphere changes dramatically from elementary school to junior high school. In the elementary schools, children are active and noisy and willing to give their opinions. Students at the junior high school level tend to be more reserved, with classrooms being less noisy.

Toshio teaches junior secondary science for three second grade classes (age 14) and one third grade class (age 15). He also has responsibility as the home room teacher for one second grade class. The lesson described takes place in a mixed ability class with 38 students.

Each class has one homeroom teacher who has responsibility for meeting them at the beginning and the end of each day. Each subject is taught by a specialized teacher. This is one of the biggest differences between elementary and junior secondary school. The homeroom teacher gives guidance and counseling to students and manages the class. This role becomes especially important in grade three when decisions are made about which senior high school students will apply to.

Almost all schools have extra-curricular activities, including sports, culture, and science clubs. Teachers have responsibility for these club activities, often making their work days very long.

The class begins as the "Nichoku" (student leader for the day) says "Kiritsu" (stand up) and "Rei" (bow). Students and teacher bow to each other. Toshio begins the lesson by checking student attendance. The topic for the lesson is a continuation of the topic from yesterday: air pressure and wind. A worksheet is passed out on the topic and students are told which page in the textbook they might also refer to. The title for the lesson is written on the board: Relative Humidity. Students take this down in their notebooks.

Toshio begins the topic by asking where the rain water goes after the rain has stopped. After waiting for a student response and not getting one, he then writes the question on the board: "Where does the rain water go?" Students are asked to work individually to make a list of possible answers to this question in their notebooks.

> *Japan has also recently revised the national course of study for junior secondary science. The objectives for science are:*
>
> *To enhance students' interest in nature, to foster scientific attitudes, to foster the ability to think scientifically through observation and experimentation, and to deepen students' understanding of natural events and phenomena.*

Students continue to work individually while Toshio walks around giving suggestions. After a few minutes have passed, he begins asking students to tell what they have written in their workbooks. Toshio then asks a second question: "Where does the water that's in the wash go when it dries?

They begin a discussion about vaporization of water and how difficult it is to observe. He then asks the class at what temperature water vaporizes. He reminds them that they have learned this in a previous science lesson. All students answered 100 degrees centigrade. He writes this on the board and proceeds to ask if this is the only temperature at which water vaporizes.

He begins to present everyday situations for the students to think about in which water vaporizes at temperatures lower than 100 degrees C. He asks them to think about how much water is contained in the air. He reminds them about the physical education class they have just come from and whether or not they perspired. He asks them, "Where did the perspiration go?"

At this point in the lesson, Toshio asks the students to open their textbooks and refers to a graph which shows the relationship between temperature and the maximum amount of water in the air. He says that vaporization of water is closely related to temperature, creating differences between a cold and a hot day. Students are now busy copying the graph into their workbooks as the teacher continues to talk.

Toshio uses the graph to ask students questions that demonstrate their ability to interpret the graph. He then gives a problem to students to calculate individually. After a few minutes, one student is chosen to go to the board to write his solution and they discuss the results.

In the next problem, Toshio asked the class to calculate the amount of water in the air in the classroom. While working with this problem, he introduces the term "humidity" for the first time. The discussion continues as he introduces different temperatures and students compute the relevant percent humidity for each temperature.

The bell suddenly rings as they are in the middle of the calculation. He tells them to memorize the equation he has written on the board and then the lesson ends as the students and teacher bow to each other.

Many students in metropolitan and suburban areas take private Juku courses after the regular school day is over. Entrance examinations for students are very difficult. The Juku courses are designed to focus on the specific knowledge and skills students need to pass these difficult entrance examinations.

The junior high school curriculum has a more academic orientation where students pay more attention to the acquisition of basic skills and knowledge, rather than on understanding and thinking. The new reform measures will try to place more emphasis on understanding, thinking, and interest in the academic subject areas.

NORWAY:
CASE STUDY

FOREWORD

Norway is a sparsely populated country with less than 4.3 million inhabitants spread over an area of about 324 000 square kilometers (13 people per sq. km.). Many of the schools in Norway are small (about half of the schools have less than 100 students) making national objectives for providing equal access to education often difficult and costly.

The Ministry of Education has the overall responsibility for administering the educational system and implementing the national curriculum. All public education is free, including tertiary education (see Bjørndal, 1995).

Norway is currently in a period of many educational reforms which will have direct consequences for the teaching of science at all levels. There will be a new emphasis on science and the environment at the primary and junior secondary levels, resulting in an increase in the time spent teaching science topics. Though the cases reported in this document do not reflect the reforms, they do provide a description of the way science is currently taught in Norway.

SCIENCE, POPULATION 1

School began in the middle of August. There was already a crispness in the air indicating the end to summer. Fall was quickly approaching as leaves began turning colors and days became noticeably shorter.

This year Kari has fourth graders. She knows the class of 24 well; she has been their teacher since first grade and will continue with them until they complete grade six. She has taught them how to read and write and how to care for each other as a group. The class has literally grown up together with Kari as their teacher. They are a good class though sometimes have a tendency to be outspoken and boisterous. With only a few exceptions their performance is homogeneous allowing her to plan lessons for the whole class. Her form of management has gradually allowed her students to assume more and more responsibility. They are clever at taking directions and working independently.

A visitor to a Norwegian primary school would see children sitting at desks, arranged in pairs or in groups of 3–4. All children have backpacks which they use each day to transport their books, supplies and lunches from school to home. Those books that aren't being used are placed in shelves, one for each student. Class size varies throughout the country from small rural schools with only a few total students, up to large city schools having 26 children in a class.

Norwegian schools are governed by a national curriculum. At the primary level (grades 1–6) science is taught as a part of an integrated subject called Orienteering, which includes science, social studies, history and geography. Teachers are often free to set up schedules for instruction of individual subjects, allowing for thematic teaching over a longer period of time. Orienteering is allocated 4–5 hours per week in grade 4.

Norwegian children are never differentiated into fast or slow groups according to their ability levels. It is not until the high school level that pathways through the educational system begin to differ from student to student. The ideal primary class is one in which most students are performing at the same level.

Keeping in mind that the same group of children will often be with the same teacher from grades 1-6, one realizes how important cooperation is in our school system. Classrooms are basically non-competitive environments where the individual is always encouraged to do his or her best. What goes on outside the classroom when children play may be a different story.

Independence and cooperation are important themes in Norwegian schooling - a primary component of its recognized social function. The "Curriculum Guidelines for Compulsory Education in Norway" (Ministry of Education and Research, 1990) explicitly makes this point:

> *It is impossible to develop a democratic attitude unless all persons are respected for their individual qualities and treated as independent individuals. At the same time it is necessary to show consideration for the needs of the community. The pupils must learn to understand that living together and sharing work and responsibility in the home, the school and the society requires cooperation. Independence and an understanding of the need for cooperation are closely related to the pupil's conception of self, and to the norms and attitudes they meet in the environment in which they grow up (p. 24).*

FALL FLOWERS

Kari begins almost every year with a unit on the change of seasons from summer to fall. This year the concentration would be on fall wild flowers. Together with the two other teachers who have fourth grade classes, she has planned to use 5 class sessions to take them out into nature to observe the late flowers of summer, to make a collection of the flowers they find, and to systematically name the flowers. How many different flowers would they find this year? How many names would they manage to learn?

Kari introduced the unit on Fall Flowers by telling her students everything they would be doing in the next two weeks. She wrote the activities on the board and allowed students to ask questions. Instructions for the next hour were given to the entire class.

Teaching is often teacher centered. The board and teacher's desk tend to be at the front of the room where teachers also lead discussions or give directions. Though children sit in groups, there is very little cooperative group activity going on. The classroom atmosphere is generally relaxed giving students the opportunity to move around and talk quietly when the situation allows.

For this project, students were to work in pairs. Each pair was to collect as many different plants as possible from the field immediately adjacent to the school. Plant roots were not to be taken. Students were asked to predict how many different flowers they might find before beginning their field work.

The class was organized to leave the classroom and walk to the field site where they had worked before on a different project. At once, the working pairs began to collect flowers and count. There was a certain sense of wanting to find the most flowers as each group followed the progress of the other groups. After 30 minutes outside, the pairs returned to the classroom to start their second science session.

The next hour was filled with anticipation as pairs began counting their specimens. Soon the entire class was involved in finding out how many different flowers they had found. Even though fall may seem like the end of the growing season in Norway, 30 different plants were found in the field site. No one had predicted that they would find so many different flowers!

The remainder of the time was spent naming the plants with the help of reference books. There were not enough books for each pair so some worked in larger groups while others waited for books to become available. Though Kari knows the names of most of the plants, she did not give out names freely to those who asked. The point of the exercise was to be comfortable using reference books as a means of identifying a plant name.

Students began drawing pictures of their plant collections in their science workbook with names written underneath. Many of the students chose to write both the Latin and the Norwegian names. The plants were placed in water until the next session.

One hour was used the following day to look at the plants with the help of a magnifying glass. Work continued on drawing the plants in the workbooks and finding out more about them from the reference books.

On the third day of the project, Kari asked students to choose one flowering plant to draw in their workbook. They were asked to draw the root, stem, leaves, and flowers. Reference books could be used to help with the details. Kari used the board to talk about the plant parts and the functions of each. A discussion on the food chain followed and how plants play an important role as primary producers. Kari also passed along interesting information about what different plants have been used for in the past. Students were asked to complete a homework assignment from their textbook which connected their field work to information about plants and food chains.

Homework is an integral part of the school experience from the first grade. In the early grades, homework becomes the link between school and home; between teacher, student, and parents. Later, as children become more independent, homework is seen as an extension of the class lesson which is to be finished at home.

Kari used the final session as a form of assessment to find out how many different plant names the children had actually learned. Pairs were asked to return to the field and pick one specimen of every plant they could name. They brought the plants back to the classroom, sorted them and then wrote the names under the plants. In the end, if they were in doubt, they were able to use the reference books. Kari made sure that the plants had been given correct names.

Formal grades are not given until grade 7, allowing teachers freedom to conduct informal assessment tasks. Student progress is discussed between parents and teachers in two conferences each school year.

Later that day, Kari integrated the science activity into an art lesson. Students were asked to draw a large picture of one of the plants they had found (as correctly as possible). The pictures were hung up on the bulletin board making a very colorful display for the duration of the semester.

IS KARI A "TYPICAL" TEACHER?

Kari is enthusiastic about teaching field experiences in science because she considers the outdoors to be a personal interest and hobby. Like most teachers at this level, Kari takes her students out of the classroom and into the surrounding environment. Knowledge about nature is a part of the Norwegian culture and is thus reflected in schools. The little science we do see taught at the primary level tends to be characterized as "Nature Study". Teachers and textbooks are careful to give names to plants and animals in the local surroundings; careful to discuss how animals and plants adapt to seasonal changes.

Science (including biology, chemistry, physics, and earth science) tends to be a very neglected subject at the primary level for a variety of reasons. Teacher education has not required science courses for the teachers of grades 1–9, resulting in a teaching force that does not feel comfortable teaching the subject. Science and civics are integrated into one subject and one textbook. The textbooks have not done an adequate job presenting science topics and teachers are free to skip over science chapters if they choose. A reform in the national curriculum will be implemented in 1997 making science a more visible part of the primary curriculum.

Kari has taken some science courses in teacher education making her a somewhat atypical teacher at the primary level. In fact, in her school she is the only teacher with a science background and is therefore considered to be the science "expert". Kari is typical because she uses the surrounding natural enviroment as a part of the curriculum. She is atypical in that she enjoys teaching science and helps her students to arrive at a greater understanding of the environment rather than simply placing names on objects. Norway hopes to have more of this type of primary science teacher in the future.

Primary schools are dominated by teachers who have degrees from 3-4 year teacher education programs at colleges of education. They have teaching credentials which allow them to teach all subjects in the com-

pulsory school (grades 1-9). Most primary teachers have little to no science education in their teacher education backgrounds. Teachers may also be qualified to teach science in grades 7-12 after completing an academic degree and taking pedagogy courses for one year. This dual system for teacher education results in two types of education for teachers in grades 7-9 where there is an overlap.

SCIENCE, POPULATION 2

This picture of teaching in Norway is framed by the participants in the classroom, the teachers and students, and the interactions that go on between them in the classroom environment. The picture has four sides, the first of which is characterized by the content represented in the aims and objectives of the topic. The second is the background, prior knowledge, experiences in the topic and the personalities of the participating students. The third side is the materials and resources that are made available for instruction by the school, and the fourth is the teacher's knowledge and his background and pedagogical beliefs that influence both the pre-active and the interactive phases of the teaching of science. Finally, is the making of the picture itself: the process of implementing a topic of science from the outline in a curriculum to the teaching in the classroom and culminating with the students' learning of the topic. This is illustrated in the following sequence of lessons.

PRESCRIPT; FRAMING

Twenty-seven teenagers are entering the science room. There are backpacks and snowy winter coats as well as semi-wet boots everywhere. There are nice words and not so nice words being said among the students about classmates, sports, parties and all the other really important matters in life. There are desks and chairs being shuffled around the room, because other users of the room rearrange everything! On a snowy winter day at the end of February it will take some time for 15 year old students to sort out their belongings, find their places and put their backpack beside their chairs. Some of them have a few words they need to say to the teacher, who is calmly waiting beside the entrance.

This situation is taken from a ninth grade class, one that we will follow through six lessons in the teaching and learning of human physiology over a period of four weeks. Nevertheless, this could be taken from any group of students in Norway the very same morning.

It is common to organize the school day in lessons of about 45 minutes and recesses of 10 minutes. Lunch break is a little longer, lasting on the average of 30 minutes. Most classes change their location several times during a school day. As a consequence of this, students normally need to bring their books and other materials with them in the same backpack that they take home in order to do their homework.

In grade 9 students typically have three 45 minute science sessions each week. It is up to the individual school how they will schedule these sessions. The ideal situation is one where there is one 45 minute session one day and the other two sessions back to back so that experiments may be run without interruption. However, scheduling problems with teachers and students do not always allow for the ideal situation.

THE JOINING OF THE FOUR SIDES

Our class is now just about ready to start a lesson. The teacher, Lars, is still patiently waiting beside his desk in front of the blackboard. Finally (10 minutes into the lesson) he asks the students to stand up and greet him the way they normally do when beginning his first lesson every day.

LESSON 1

Lars spends the first few minutes explaining the planned activities for the lesson. His intention for the lesson is to measure the oxygen intake of the human body. The students start immediately to work in two subgroups. Half of the class works with written tasks taken from the workbook and the other half carry out a step test. They work mainly in pairs, sharing the tasks so that one student is exercising and the other student recording time. They use a weight, a watch, a diagram and a low bench. They follow written instructions, but many of the students seem to be absorbed in aspects of the topic which relate to sports. Discussion about the activity, among the students and between the students and the teacher, involves therefore different kinds of sport games and how they vary with respect to inhaling, exhaling and heart rate. After completing the activity they return to their desks and start the written assignment that concludes the activity. To solve the tasks from the workbook they use their textbook as well as communication with classmates.

Lars acts in this lesson as a typical facilitator, he organizes the activities in advance and during the lesson he hands over the control and progress of the activity to the students to a large extent. Another main feature of his role as a teacher in this lesson is that he quite often answers questions by directing another question to the students trying to challenge their way of thinking about the subject.

Students are quite motivated for this kind of experiment. They ask lots of questions. Some of the questions are procedural questions which shows that it is important for them to obtain correct results and they ask for advice to get on the right track. Other questions are related to the understanding of the main issues and conclusions that can be drawn from the results.

LESSON 2

The next lesson starts where the first one ended, with an extended discussion about results from the step test and implications of these both for human physiology and physical activities. The main body of the lesson is a presentation of the theory of human physiology that supports the results of the step test. Among the details that are presented are the composition of air, lung capacity, smoking and inhaling of smoke, lung cancer and its causes and the anatomy of the respiratory passage and alveolae. The presentation of different people's lung capacity, smokers and non-smokers, physically active and physically inactive, is done by the direct use of the students in the class, Lars and a teacher that, by coincidence, enters the room. Towards the end of the lesson the teacher brings out some equipment to demonstrate inhaling and exhaling. The students are at this time occupied with the remaining written tasks from the preceding lesson and no one takes up his invitation to demonstrate the equipment.

The teacher conducts the main body of the theory presentation in this lesson and illustrates by drawing diagrams and charts on the blackboard. His procedure is, nevertheless, built on the students' experiences with the last lesson(s) and most of the details are filled in by the students at the teacher's request. The students are quite eager to give him input and he manages to weave in most of their replies either by a direct use of the answer or by following more closely the statements given. Several times during the lesson there are question and answer interchanges between Lars and individual students. The illustration on the blackboard evolves as the discussion continues and at the end of the sequence there is a structured and complete representation on the blackboard.

During this lesson the students seem to be very motivated to talk about cancer and other diseases, their causes and implications for living a normal life.

Sometimes the teacher is caught between letting the students build on their own associations as a motivating factor and keeping the conversation strictly to the subject in order to fulfill the intentions of the lesson. On the one hand, Lars strongly communicates his willingness to pick up almost any question or remark posed by the students, but on the other hand he wants to bring the students to a certain point of comprehension with the lesson's topic.

LESSON 3

A new activity is introduced in this lesson. Today the students will be typing their blood for blood groups. Before the lab work they discuss safety and the implications of blood work in connection with contagious diseases like Hepatitis A, B or C and AIDS. In addition, some of the concepts from the last lesson's "theory presentation" are reviewed and a few more, like air pollution, blood system and "respiration" are added. As a second introduction to the activity the teacher has a quick presentation of different blood groups, donation of blood and blood coagulation. This review and discussion leaves the students with about 15 minutes to do the lab work.

As opposed to the execution of most activities in this class the students have problems starting the practical work and they need to be pushed. They seem to have their attention directed at other things, and they also have severe problems in drawing conclusions based on the work they do. Lars is aware of this situation and the complications for the students. He moves very quickly from student to student to answer questions and give directions about the activity. He has to explain the concept of antibody several times before they assimilate this and can transform the knowledge and anticipate its significance for blood typing. Due to the fact that the group, in this lesson, consists of only 14 students (the class is divided in half for one of the three lessons each week) most of them manage to finish the task before the bell rings.

LESSON 4

The teacher uses this lesson to present information about the functions and anatomy of the human heart and the blood system. Lars explains that this is necessary information for the coming dissection of the heart and lungs from a pig. They then return to the explanation of the blood typing activity they did in last lesson. In addition, they have a quick round among the students to sum up the replies of the written tasks they have been working on at home and in class between other activities. When going through the blood system the teacher makes use of the blackboard for illustrations and the students' knowledge about details. The students are also asked to take down notes in their workbooks. They listen to a tape recording of a heart beating. The lesson finishes with an activity of listening to each other's heartbeat when sitting still and, again, after running around the building.

It is clear from the students' statements in the first part of the lesson that, despite the messy circumstances of the exercise, they managed to draw the right conclusions from typing their own blood. The students demonstrate interest for the subject by being very eager to give answers both for the written assignment and for questions posed during the presentation at the blackboard. Their reward for their collaboration is the heart and lung dissection which concludes this topic. Even though the class is cooperative during the presentation of theory, there is no doubt about their preference for lab work and other activities. Concepts are learned through a combination of different activities in the class. Lars underlines the motivating factor of lab work and variation in teaching activities.

LESSONS 5 AND 6

The classroom is prepared for the dissection of the pig's heart and lungs. Lars has organized the students' desks in a half circle and there are two larger tables in the middle. The students start dissecting after having been divided into two groups. They follow a sheet with written instructions at the beginning. The teacher planned this activity to take one lesson but at the end of one lesson realized that a second was needed (not on the same day). The first lesson, therefore, involves the dissection of the lungs. They look at the lungs under a dissecting microscope and weigh the heart. In the second lesson, during which they dissect the heart, they work in pairs. Lars gives detailed instructions showing them how to hold the heart, how to cut, and what to look for. In the small working groups he explains how the heart works and later repeats this for the whole class.

After some input from the teacher, students start experimenting on their own. They find better solutions when it comes to task sharing and generate hypothesis they find interesting to explore. The teacher encourages this process. Lars is very generous about mixing fun and lab work in the first lesson while in the second lesson he is more concerned about the heart dissection and the need for a certain degree of accuracy in the work to see the details of the heart's anatomy.

After the six lessons, he assumed that there were some aspects of this topic that had not been covered. When asked what remained of the topic he replied:

I intend to go through the function of the heart one more time, using the illustrations they have. Then they will write a lab report entitled, 'Murdering the pig'. They once wrote a detective story based on the classification of chemicals, and I intend to repeat that type of report since they liked it. There is no reason for always doing lab reports in the same style.

POSTSCRIPT; THE PAINTING

Twenty-seven teenagers are exiting the room. They have their backpacks ready. Their winter coats and mittens are back on. Some of them have blood typing, respiration or other "science words" in their minds but from the look on their faces it is clear that many of them have throwing snowballs, girl-friends and boyfriends, and maybe the next subject in the back of their heads. Their heartbeat is with them. They will always feel it but not always count and measure it. They all say good-bye to Lars, and some of them have comments about the lesson.

Our teacher, as well as the other teachers in Norway, has a national curriculum guideline to guide the planning of teaching sequences. In addition, the national curriculum is intended to be adjusted to the local circumstances. Hence, Lars has a local (municipality) adapted plan containing suggestions for science teaching based on the local priorities for operationalizing aims and objectives, the methods to be used in teaching, and the locally available resources and offerings. Our teacher, and others with him, are of the opinion that a process of educational planning based on this national curriculum is an important stimulus to illuminate their pedagogical consciousness and reflection about the con-tents and concepts of science in order to improve their teaching and the students' learning.

All school textbooks are nationally approved by the government, with several titles available from publishers in each subject and level. At the primary level, teachers tend to be very textbook dependent. At the junior and senior secondary levels, where teachers have a better academic training for the few subjects they teach, they tend to bring their own expertise into the classroom and use the textbook as a supplement. Since all textbooks are an interpretation of the national curriculum, they tend to influence the choice of topics and the order for teaching topics in all subjects.

SPAIN:
CASE STUDY

FOREWORD

During the last five years, Spain (which has a population of approxi-mately 39 million) has been involved in a reform of the educational sys-tem which has affected both the organization of the grade levels and the curriculum (Ministerio de Educación y Ciencia, 1989). In the syllabus of this reform, science is viewed more empirically in order to favor a change in the preconceptions, assumptions, and beliefs of the students and to enforce the acquisition of concepts, the mastery of procedures, and the development of attitudes like curiosity and interest in their envi-ronment. To achieve this, the new curriculum stresses the importance of these three aspects of contents (concepts, procedures and attitudes) in all subjects.

Most teachers think that science teaching reform uses experimental and laboratory work as a fundamental reference. Nevertheless, in Spain there has been a tradition of purely academic or lecture–style teaching and it is difficult to change the curriculum actually implemented in the classroom. Sometimes, there is also a lack of either adequate prepara-tion or materials and facilities that are necessary in order to execute a more practical curriculum.

More specifically, in science the general objectives to be achieved with the new curricula are: that students acquire abilities such as better com-prehension of the world through the acquisition of procedures and strate-gies to explore reality, permitting a development of comprehensive abil-ities and correct interpretation of scientific and technological texts; and secondly, that students acquire knowledge of the characteristics, possi-bilities and limitations of their body whose welfare and balance depends on its relationship with the environment. A final objective is that students develop attitudes such as flexibility, critical reasoning and intellectual rigor.

SCIENCE, POPULATION 2

It is twenty minutes past nine in the morning. Juan goes to the school that is located 200 meters from his home. He is 13 years old, as are most of his schoolmates. He walks quickly to catch up with two of them. Their school is one of the six state schools in the community. It was built in 1981. It has four independent buildings with 4,130 square meters and 5,974 square meters of open area. It has four pedestrian entrances, a football field, an athletic track, a small garden and a 5,000 square meters playground.

Two of the four buildings are similar, both having classrooms for students aged six to fourteen. There is also a pavilion for preschool teaching and another one for common service facilities. All of them have a store room for materials, a cleaning room and restrooms on every floor. The service pavilion, which is the central building, has an administrative area (housing the headmistress, secretary and teacher's room) and an education area (housing the library, laboratory, orientation department, and dining hall). The school is divided into 22 units, 16 from 1st to 8th grades and 6 units of preschool (three-, four-, and five-year-olds). Each grade has about 50 students, 25 per unit. Grades 7 and 8 will form the first cycle of compulsory secondary education under the new regulations of the educational reform.

Juan's class is located on the second floor of one of the twin pavilions. At this time of the morning the students are arriving at the school and all of them, regardless of class or age, congregate in the playground waiting for the bell to signal them to go to class. Juan has arrived only a few minutes early but still has time to play marbles with his schoolmates in the playground.

The bell rings at 9:30. Once the janitor has opened the door of the pavilion, Juan and his friends go to their class chatting enthusiastically. Once inside the classroom, some students continue chatting while others go to their respective seats. Juan has hung his coat up in the wardrobe. Once he is seated he gets ready to take out the books from his backpack.

The classroom measures 7x9 meters. On the left side, there are three windows that let in a lot of sunlight. Juan is sitting close to the windows in the third row. The room is not particularly decorated.

There are 29 students in Juan's seventh grade classroom, 16 of them are males. Three of them have repeated some grades before this. There are no handicapped students in this group. Juan's class is heterogeneous academically since they have been in the same group since preschool. The criteria for grouping students in their first year is more or less alphabetical, trying to have the same proportion of boys and girls in each group. Normally no change is made unless a problem appears.

At the beginning of the course, Juan and his friends chose where they wanted to sit and with whom. As the course goes on, and normally after evaluations, the teacher may decide if changes need to be made. At this age, the students normally choose to sit with schoolmates of the same sex. The tables are placed in rows of two which is the most typical arrangement. Nevertheless, it is not unusual for the teacher to ask them to move tables, depending on the type of work they are going to do. Sometimes they are asked to work individually and at other times in groups of four or six.

Juan always brings his backpack full of books. He has a notebook for each subject and now he has taken out the science notebook and textbook. The other children do the same and, while they wait for the teacher to arrive, they chat about the Sunday soccer match.

It is 9:40 and the teacher comes into the classroom. The students go to their seats slowly and eventually there is silence. The teacher, Manuel, is forty years old. He earned his degree after three years of studies at the University as a specialist in science and mathematics. Thus, he is allowed to teach at both the primary and compulsory secondary education level. However, most of his in–service teacher training during the last few years has been in mathematics.

Once the reform is completely implemented by 1999, teachers of secondary education (both compulsory and non-compulsory cycles, e.g., 12–16 and 16–18 years old) will need a five year university degree. Teachers at both the infant and primary education level will need a three year university degree.

For 11 years Manuel has had a temporary post in different schools of Madrid's Autonomous Community. He has been teaching various grades and subjects. Two years ago, he got a permanent post in this school and he teaches mathematics and science in compulsory secondary education with 21 teaching sessions per week. He is also the tutor of the seventh grade group which is the object of the observation.

There are 29 teachers in this school: 22 of them have a permanent post, which gives stability to the school; 26 of the teachers are women. The Headmistress, who has held this post for ten years, is assisted by a Director of Studies and a Secretary, both of whom are teachers that are released a few hours per week (three to six) from teaching. At the top of the school organization, the school board has representation from teachers, parents, students, and local educational authorities.

The academic structure is based on education cycles, two grades each, rather than on subject departments. However, in compulsory secondary education, due to both the relevance of subject matter and the special-ization of teachers, departments would be a common form of organiza-tion in some schools. In this school, mathematics and science teachers cooperate in the planning of content and scheduling of lessons.

In his notebook, Juan has written the outline of the didactic unit, "Fluids: liquids and gases", which the teacher provided to students at the beginning of the unit. Before starting each unit, the two science teachers for this course col-laborate, deciding objectives, contents, activities, and scheduling. Juan and his schoolmates use this outline as a guide, even though the teacher can always make adjustments to it taking into account any difficulties which arise.

We are in the eighth session of a 10 session sequence. Juan has science class three days a week. In the last class, Juan and his schoolmates had a very good time because they were in the laboratory working in groups of six. During the laboratory session the teacher stated a problem which demanded the manipulation of different objects rather than a paper and pencil problem.

The laboratory is situated in the central pavilion. Normally it is closed, with access only when a teacher is present. It is well–equipped. It has a small entrance with a cupboard for science books and videotapes. Inside the labo-ratory there are more cupboards with materials such as test tubes, scales, weights, dynamometers, electrical materials, etc. There is also a locked cup-board that contains two magnetometers, different sets of boxes containing materials and instructions for experiments, and a big TV screen and videotape recorder. There is also a basin and on one side there is a shelf for working which has a big set of scales. There is a blackboard on one side and six tables with stools in the middle of the room for working in groups. The walls are covered with posters about different natural science subjects.

The content of this unit has been in basic concepts: pressure and the Pascal and Archimedes principles. Last week they worked on liquid pressure which used up quite a few of the class sessions for the unit. At the beginning of the unit the teacher asked them to do a practical exercise from the textbook. Such activities are done at home and are always shown in class after the theoretical explanations. Some teams have already displayed their results to the rest of the group.

Juan's team performed an experiment to explain the communicant vessels principle. The whole team is standing in front of the blackboard and one of them is in charge of explaining to the whole class what has been done, the materials used (plastic bottles, rubber tubes and adhesive tape, etc.), and what they were trying to demonstrate. The teacher intervenes occasionally to relate theory to practice and asks for explanations and definitions.

Juan and his classmates use about ten minutes to explain their experiment. Once they sit down the teacher calls on the last group to explain their work to the rest of students. The atmosphere is relaxed and the students pay attention to their schoolmates. They occasionally make comments about the experiments they are being shown but what is more common is that the same students who are presenting the work answer the teacher's questions.

It is 10 a.m. Manuel, the teacher, asks the students to open their books to page 47 and read the section "Pressure in gases" silently. It is the last day of theory. Five minutes later, the teacher asks the whole class questions to find out if they have understood what they have read. In fifteen minutes the teacher completes and clarifies the concepts described in the textbook. In this explanation he uses the answers of the students to introduce and review concepts and to correct mistakes. As the unit progresses, the type of questions the teacher asks students demands more elaboration. He has to compare, make inferences, and even to evaluate answers from other students. Sometimes he divides students according to their point of view to create disagreements and stimulate student interest.

The teacher forces the students to justify their answers. Comprehension and practical usefulness is more highly valued by the teacher than a memorization approach to the concepts. "First to see and after to justify theoretically". During the question and answer session he explains how gases behave. Graphical examples are used to explain. He draws different images on the blackboard, for example a room full of tennis balls, or he describes how an aerostatic ball works. He suggests as an activity the construction of a ball, indicating the necessary materials and the way to construct it, and also the physical principle on which the construction is based. They have to do it in groups after school time. Next week they will go to the countryside to make the aerostatic balls fly.

Finally, he gives definitions and asks the students to write them down in their notebooks. Afterwards he summarizes the data collection content using questions that require more complex information and thought.

Juan is a little tired and looks at his watch. In only ten minutes the science class will be over. The teacher has finished his explanation and dictates two exercises to be done in class. Juan starts to do the first exercise with his neighbor. It is 10:30 and the bell rings. Carlos and Juan look at each other. They have solved only the first problem. Before the teacher leaves the class, he says: "Those that have not finished can do it at home and we will correct it tomorrow. Read the whole unit again and we will go over it tomorrow, too and clear up any doubts. Remember that tomorrow is the last day before the exam."

Juan must study because Manuel takes into account notebooks, the class marks and the examination for the final evaluation. He does not review every notebook, just those from the students with more difficulties, as a way to make them complete the notebooks every day. The classroom marks are obtained mainly through the practical demonstrations. The examination is a copy sheet with 10 questions or problems. During the exam the students work individually at separate tables. They may consult the textbook but not their notebooks. Students are permitted to use calculators.

During this unit the teacher has given information about marks. In the first session he commented on the marks of the last exam and evaluation. He does not evaluate every one equally as he takes into account their abilities. In the past evaluation just one student failed. The teacher's criteria for evaluation has been that every student has to know the key concepts and their practical application in daily life.

The Educational Authority establishes the objectives to accomplish by the end of each grade and proposes some evaluation criteria. Following those regulations, teachers have a considerable amount of freedom to establish their own criteria. There are five "Evaluations" throughout the course and teachers have an evaluation session before each one, making comments about the students' results in all subjects. After that, depending on the routines of the school, the teacher sends parents a written report or just a bulletin with the evaluations. This school uses a computerized sheet to inform parents.

The teacher leaves the classroom. Juan puts his science textbook and notebook away in his backpack. He is a little bit worried because he knows he has to study this afternoon and he tells this to his friends. The students continue to chat while they wait for the language teacher to arrive, which she does four minutes later. It is a new class, and new material is taken out of Juan's backpack.

MATHEMATICS, POPULATION 2

It is 9:15 on Wednesday morning. Fifteen minutes have passed since the beginning of the school day. The first hour, as usual, is dedicated to mathematics. If we were observing the class through a window we would see a group of 27 students seated in groups of two, a blackboard and the teacher's table at the front of the room. The students are working individually in their notebooks even though we can see some of them consulting or chatting to their nearest schoolmates; there are also some students who are moving around, either going to the wastepaper basket or consulting the teacher. The teacher is moving along the rows observing the students' work, answering their questions and, sometimes, making suggestions.

What is the topic of the day? If we look at the blackboard we can see several coordinate axes with points represented on them. Most of the data consist of rational numbers. This corresponds to the activity they did during the first 15 minutes of class, which was correcting two exercises the teacher had given to them as homework. As you surely have guessed the topic they are working on is "Functions". The teacher started this didactic unit two days ago and she wants to dedicate one and a half weeks to it –six sessions– as she agreed to do with the other teachers in the department when they planned the course activities.

In Spain, with a new Act reforming the education system (LOGSE, 1990), the education administration of each Autonomous Community establishes the basic curriculum for each education level (infant, primary, and secondary) taking into account the obligatory core curriculum for the whole country. The basic curriculum specifies the general educational objectives as well as the general content (concepts, procedures, and attitudes) and assessment criteria for all the different levels and subjects. These basics of the curriculum are stated explicitly for teachers in the School Curriculum Projects, adapting objectives and content to the concrete reality and needs of students. Classroom teachers then devise their own program taking into account the decisions made together with others teachers in the department.

When the teacher presented the unit to the students last Monday, she checked on their previous knowledge of the topic. They reviewed what they had learned last year. She wrote some key vocabulary items on the blackboard, like the Cartesian coordinates, abscissa and ordinates, etc., and she asked what they meant and how to represent various points. The students answered the questions voluntarily although sometimes, to confirm or modify some answers, she asked individual students. Thus the teacher and students reviewed content that had already been learned, generating a common vocabulary to be understood by the whole class.

What are the students doing now? Let's go up to one of them. We can see on the table a ruler, one blue pen and another red, a pencil, a notebook and a textbook opened at the same page as the rest of the class. The students are copying an exercise and solving it in their notebooks.

The textbook is an indispensable resource for most Spanish teachers. The majority of students have their own book. Teachers normally follow the guidelines in the textbooks. The contents are organized into themes and generally include corresponding exercises. Sometimes teachers supplement the textbook with other materials or textbooks from different publishers, of which there are six or seven. When the teachers prepare the didactic unit they go to several textbooks which they may have at home or from the school library.

The students themselves have textbooks which they can consult in the classroom library. In this class the library is located near the teacher's desk. There are more than seventy textbooks including encyclopedias, dictionaries, atlases, other student textbooks and children's literacy books. On the other side of the classroom we can see a cupboard containing materials like bristle boards of different colors; scissors; squares and compasses for the blackboard; an audio tape recorder; glue tubes; newspapers; adhesive tape; past reviews; and some wooden pieces. We can also see some geometric figures that were made in a previous unit. The classroom has cork panels all around. One of them is devoted to organizational matters, like rules of living together, schedules, reading assignments, number of books read by each student, lost items, etc. Another panel is dedicated to relevant news – at the moment, ecology, environmental preservation, and drought. There are also maps and students' schoolwork.

Our students continue working in their notebooks. Some of them sometimes require the teacher's attention, asking questions in a loud voice. The teacher gives instructions and responds positively when students' work is good. She corrects mistakes and when they are repeated by several students she uses them as examples for the whole class. The teacher helps students individually and students also talk between themselves and help each other. At the moment the teacher is helping two students at the back of the classroom. Suddenly she looks up and asks for silence, since the level of noise has increased considerably. Generally speaking, the students are allowed to talk if they do not disturb the other students' work. They are also allowed to get up from their desks to ask the teacher or other students questions, as long as there are not too many standing at the same time. We know that students can choose where and who they want to sit with at the beginning of the course. They can continue to sit with whom they like during the course except when discipline problems or comprehension difficulties arise. In such cases the student is made to sit near the teacher's desk or close to the blackboard or sometimes he or she is asked to sit with a brighter student.

Taking advantage of the working atmosphere that has been restored, we look out of the window. We see a playground with students from another class doing gymnastics. Several trees are part of the view, as is the main building of the school where school administration and the primary courses take place. The type of building tells us we are in a private school, since it consists of two chalets turned into a school.

There are public and private schools in Spain. Normally there is more uniformity in the construction of the public schools than private schools where diversity is more common. Some private schools are new constructions while others are modifications of existing structures. Both must abide by the statutory provisions.

Private schools are sometimes sponsored by the government. They can be religious or secular. Most of the school population attends either a public or sponsored private school both of which follow a similar curriculum. The biggest differences may be found in the educational ideology, also in some of the extra–curricular activities and, as happens at any school, in the personal style of the teachers.

Nevertheless, as far as the organization of the school is concerned, there are two basic differences between public and private schools. First, is the selection of teachers. In public schools the teachers are civil servants who have passed an official examination which, on one hand, gives them a post for life and, on the other hand, obliges them to work during their first years as a teacher, in different towns, schools and courses, until they are given a permanent placement. In private schools, on the contrary, the teacher is taken on by a particular school and given a private contract. Secondly, public schools are directed by one of the teachers democratically elected by his or her colleagues and for a period of no more than three years. In private schools the director is elected by the owner of the school and he very often stays for several years in the post.

Coming back to our class we ask Rosa, the teacher, about her work experience. She told us that she had been teaching for fifteen years in the same school, twelve of which have been teaching mathematics. She is also a pedagogue and participates actively in the curricular decisions of the school as coordinator of the mathematics department. Outside the school she works in a mathematics teachers' group. They investigate and study different mathematical issues and didactic matters.

It is 9:35 and Rosa informs the students that they have to finish the exercises they are doing so that they can be corrected. At the beginning of the class she usually tells the students the tasks they will be doing during the session and she controls the time they do it in. Rosa asks one student to go to the blackboard to solve one of the given exercises.

As the student writes the answer her schoolmates express their opinion of it loudly. The teacher realizes that several students have made the same mistakes and then, she realizes that the students have not understood the topic properly. She is worried because she knows she has to give more time than she had planned to increase their understanding and that will delay the overall program. We know that a lot of teachers are worried because they need to cover a curriculum which is sometimes too broad to allow them to go into the contents more deeply. In this case the problem raised is to represent the graphs to scale and correctly depict points.

Once the doubts are cleared up, and when there is only ten minutes left of the class, Rosa decides to present the class with new material. As she cleans the blackboard she comments that the exercises which have not yet been looked at will be corrected by her and that the students have to give her the notebooks at the end of the class. The results of the exercises will be taken into consideration when she makes a final evaluation of their work. Some begin to protest because as they point out they have only done the exercises in draft form. The teacher gives them the alternative of presenting the notebook the next day.

During the next few minutes the teacher introduces the concepts of independent and dependent variables. She explains to the students how to represent them in a Cartesian plane. She gives an example from real life. In this way she gets the attention of the whole group.

The bell rings and brings the class to an end and Rosa tells the students that they will continue the topic tomorrow. She asks them to think of two more examples about variables for tomorrow's class. Several students move near the teacher, some to give her their notebooks, others to ask questions about the next excursion, or the examination date or their marks from previous home-work. Rosa takes the books from the table and goes to the door where Ramón, the history teacher, is already waiting.

It is 10 a.m. Today, Rosa does not have any spare time. She still has four more classes, two in the morning and two in the afternoon, as do her students. The four sessions are separated by a long break for lunch.

SWITZERLAND:
Case Study

FOREWORD

Switzerland is a small country of about 7 million inhabitants. Living in a neutral, multilingual and multi-cultural country in the heart of Europe, Swiss people are still proud of their independence. The direct democracy, as well as the political independence, are traditional and highly emphasized values. Switzerland is made up of 26 democratic and independent cantons. These cantons still differ according to languages (German, French, Italian, Romantsch) and, originally, also according to Christian denomination, e.g., Protestant or Catholic (Federal Statistical Office, 1991).

The federalist union makes it hard to speak about Switzerland as one country. What is true for the political system of our country is also valid for the Swiss school system: the principle of federalism. Centralism is not welcome. No National Department for Education exists. Cantons have their own school systems and school laws. Some of them are very similar; others are quite different. For instance, compulsory attendance at school comprises nine years throughout the country, but the structure of the school system during this period differs considerably from canton to canton. In some cantons primary school takes up four years. Then there follows a process of selection. About half of the children are selected for a school with lower demands, while the other half are selected for a school with higher demands at the lower secondary level. In other cantons, primary education may last five, six or seven years up to the first process of selection. There is also a diversity at the lower secondary level; there are two, three or four types of schools with more or less exacting standards. This diversity concerns the structure of a school as well as the structure inside the school. Every canton develops its own curriculum, and most of the larger cantons publish their own textbooks (Gretler, 1995).

Post-compulsory education is more standardized, particularly because the Confederation governs laws about vocational training and sets entry qualifications for the two Federal Institutes of Technology and for medical studies. In this way, the Confederation manages to standardize the syllabuses of cantonal schools preparing students for the University Entrance Certificate (Matura).

The term "school" used in Switzerland connotes diversity because numerous small schools with one, two or few classes/teachers exist, especially at the primary level. A school might comprise a house with one classroom, one teacher and twenty students in different grades. Most schools at the primary level are so small that they have no administrative personnel. The function of the headmaster is taken by one of the teachers. The school is only an organization. Its size depends on the numbers of students who live in the neighborhood (Moser, 1993).

MATHEMATICS, POPULATION 2

SCHOOL LOCATION AND ORGANIZATION

The "Dominik School" is located in a wealthy quarter of Bern, the capital of Switzerland. Bern is a rather small city with 135,000 inhabitants. The school's building is big and modern, radiating a friendly and pleasant atmosphere. A small park with animals helps create this atmosphere.

The "Dominik School" is a rather large school with about 600 students in grades 1–9. The headmasters' job, undertaken by one of the teachers, is divided up into 60% leading the school and 40% teaching. There is another half-time secretarial position.

This school provides secondary education. In the Bern canton, students begin secondary education after four years of common primary school. Secondary school (grades 5–9) is divided into three different levels. The "Dominik School" is a comprehensive school which means that the students can easily change from one level or track with the education system to another. The described lesson was observed in a class of students at the highest level. After four years of secondary school, students of this level usually continue their further education at the gymnasium.

CLASSROOM

The size of the classroom is about 8 x 6.5 meters. It is a room full of light from the windows that constitute one entire side of the room. It is arranged in a rather traditional way, the teacher's desk stands in front of three rows of students' desks which are directed to the teacher and the blackboard. Two students sit at each desk. The room gives a rather impersonal yet tidy impression. The walls are bare: no posters, plants or other personal things of students decorate them. A couch is placed on the right of the blackboard. The blackboard in front and an overhead projector beside the teacher's desk serve as educational media resource.

TEACHER AND STUDENTS

The class consists of 11 girls and 12 boys all between the ages of thirteen and fifteen years. They give a friendly impression. The percentage of foreigners in the class is below the average for Bern. This is the students' second lesson of this morning; the preceding one was a French lesson. The students are now confronted with a double lesson of mathematics from 10:00 until the lunch break. It's striking that girls and boys don't mix; either two boys or two girls sit together at one desk.

GENERAL CONTENT OF THE LESSON

The observed lesson deals with geometry: the "middle-perpendicular" and the "bisector of an angle." Students are expected to solve corresponding problems including the construction of these special lines. These topics are discussed in the introductory chapter of basic understandings and constructions in geometry. The students deal with some basic notions, ideas and figures; they are expected to know and apply them in problem situations. Furthermore, the introductory chapter includes some prescribed constructions and, in a more general sense, the comprehension of geometry or pictures as the abstraction of a designed reality of our environment.

The students of this class have had geometry lessons for just a few weeks, yet they have already acquired a basic knowledge of the subject. The meaning and characteristics of a "middle-perpendicular" and "bisector of an angle" are expected to be clear and they have learned methods to construct them independently. In the previous lesson, the teacher required them to solve some exercises as homework.

LESSON ORGANIZATION AND EDUCATIONAL METHODS

Usually, a lesson lasts for 45 minutes. In this case, the teacher has divided this period into three parts using different kinds of educational methods and class organization:

In the first part (20 minutes), the teacher goes over the homework exercises. For this educational unit, he uses the blackboard to demonstrate some solution-steps. There is an interchange between teacher and students; the teacher asks the students, and the students give answers.

In the second part (20 minutes), students start working individually. They get a worksheet with three exercises containing geometric problems. They are asked to solve them independently, but working with a neighbor is allowed. The teacher answers questions if anyone needs help.

In the third part (5 minutes), the teacher demonstrates an example of a more complex problem. For this, he uses the overhead projector and the blackboard. The solution method is elaborated in a discussion between teacher and students.

LESSON PROGRESSION

Students enter the classroom and take their seats. As the teacher wants to start his lesson, he asks for silence, just once. The students are expected to have solved some exercises as homework. The teacher does not walk around the desks checking the solutions of every student. He begins to call on the students (whether or not they put their hands up) and demonstrates some solution-steps on the blackboard using dividers and a ruler. Most of the problems are purely geometric constructions regarding perpendiculars.

Students review and, if necessary, correct their solutions. If they are asked questions, their answers are rather short; one or two words only. The teacher often expands and explains their statements. Only once does a student explain a construction using more than five sentences. Some of the students ask questions (for example, "What proof corroborates a statement about intersection?"). In this first part of the lesson, the students participate actively and the atmosphere in the classroom seems relaxed.

After he has finished reviewing the homework, the teacher distributed a worksheet containing two problems (taken from the mathematics book used in seventh grade, secondary school – Geometry I). The students are asked to work in groups of two, solving these two exercises independently. Discussion in a soft voice is permitted; the teacher can be asked questions if necessary. Shortly after beginning individual work, a boy comes to the teacher asking him for help. They talk together for some minutes. Other students who need help, too, have to wait in the meantime. Afterwards, the teacher walks from desk to desk checking on how the students are doing, answers questions and gives explanations.

There is some noise in the room but most of the students seem to discuss the exercises. There is only a little talking about non-math topics. Suddenly, widespread confusion arises because in one of the exercises, constructions lead to a different solution than expected. In one of the last lessons, a similar exercise has been solved, but the figure concerned was a square, not a rectangle. One student gets so angry about her confusion that she doesn't solve this nor any other problem during the rest of the class.

The teacher allows students to work in groups until most have solved the two exercises and, before switching to another problem, he gives some last explanations using the overhead projector. Then he begins a new exercise, taken from their math textbook All students have a textbook on their desk. One student has to read the question aloud and the other students are expected to elaborate on the various solution steps.

Again, there is an exchange between teacher and students. Some students suggest parts of solution steps while the teacher develops the construction step by step on the blackboard. Generally, students are still attentive and they participate but the murmuring in the classroom rises and some students talk together, probably about non-geometric matters.

As the schoolbell rings, the teacher has completed the constructions on the blackboard but he still waits for a description of the characteristics of the obtained solution. He intends to deal with further questions in the second mathematics lesson and allows the students to have a five minute break.

SCIENCE, POPULATION 2

CLASSROOM

The science lesson takes place in a special room, a type of laboratory. It is bigger than a normal classroom (7x10 meters) and functionally well equipped with a TV; lots of cupboards containing various materials and chemicals; and a number of specialized laboratory installations. Four long desks for the students, one behind the other, are placed in the middle of the room; one desk can be occupied by five students. In front stands a big demonstration desk with a sink installed at one side. On the desk, the teacher has prepared an arrangement for some demonstration experiments. The blackboard, an overhead projector and a large table of the elements serve as teaching resources.

TEACHER AND STUDENTS

The class consists of 11 girls and 12 boys all between the ages of thirteen and fifteen years. For the students, this lesson is the second of this morning; the preceding one was, for some, a Latin lesson and for others an Italian lesson. Students enter the room and take their seats. All the girls sit at the two big desks in the back, the boys at the desks in front of the teacher's desk.

GENERAL CONTENT OF THE LESSON

In this lesson, the teacher addresses an introductory chapter of chemistry: "The categorization of substances and methods of their separation." These topics are discussed in connection with a latter chapter concerning "substances and their characteristics." The students get basic information about the chemical structure of various substances. They should be aware of differences in substances' characteristics and, based on substances' composition, they are expected to know existing separation methods used in chemistry.

LESSON ORGANIZATION AND EDUCATIONAL METHODS

During the observed lesson, the teacher changes his educational methods and instructional techniques very often. Beginning with an exchange with the students, he switches several times from individual work with worksheets or the chemistry textbook to discussion – asking the students questions; giving explanations at the blackboard, overhead projector or table of elements; and demonstrating experiments at the teacher-desk. In 45 minutes, he presents an enormous amount of information to the students using fairly complex concepts. Students never have an opportunity to choose what they want to work on or how to work; they must follow the clear instructions of the teacher. The students' main activity is observing, giving short answers and filling out the worksheet.

LESSON PROGRESSION

Students take their seats. The teacher greets them and introduces the observer, sitting behind the class. He briefly explains the observer's purpose and then begins the lesson. Students had to read a text for today about separation methods for substances. As a review, the teacher asks short questions about this text, e.g., "What is crystallization, filtration, sublimation?" The students are attentive, often raise their hands to respond to the teacher's questions but give rather short answers. So in many cases, the teacher comments on their statements and gives further explanation or more detail.

After five minutes of this type of discussion, the teacher switches to individual work. He distributes a worksheet and asks the students to read it carefully before answering the questions. They are allowed to work with a partner and to discuss the questions together. Students begin working. They are still very attentive and everyone seems to be busy completing the worksheet. During this period, the teacher stays at his desk looking around from time to time to see if anyone has a question. One student comes to the teacher and they briefly talk about a problem on the worksheet.

After a few minutes, although not all students have yet finished their worksheet, the teacher begins a discussion of the worksheet with the entire class. Again, he asks questions and students actively participate. The teacher collects answers and gives further explanations or examples from everyday life (e.g., "milk is a heterogeneous mixture"). As teaching resources, the teacher uses the blackboard to write some notions or explanations and the table of elements to show some details.

The topic changes from general concepts to specific separation methods. The teacher asks the students to take their textbooks and to open them to page 90. One student has to read aloud a passage about chromatography. As the passage is rather long, the teacher asks another student to read aloud, too.

Now the teacher begins two demonstrations at his teacher-desk. He asks the students to settle round his desk, so that everyone can follow the demonstrations. First, he shows how chromatography proceeds. For this purpose, he takes a round blotting paper on which he has applied a spot of black color in the middle. With a knife, he drills a hole through the color-spot and puts another blotting paper, formed like a little tube, through it. Now the blotting paper is laid on a petri dish filled with water. After a period of time, the water diffuses from the paper-tube over to the round blotting paper and so steadily spreads from its middle to its margin. The different color substances are transported by the water but each color travels a different distance. In this way the different colors are finally separated and form colored rings around the middle.

While demonstrating, the teacher explains each step. After the experiment is completed, he asks questions about the results. The students are still participating well; some of them even ask for more precise information about the demonstration.

The teacher begins the second demonstration, a distillation. For this procedure, he has prepared all the needed instruments in advance. He begins to heat the alcohol (wine) with a Bunsen burner. He explains the different processes briefly and then lets the students go back to their seats because this experiment needs some time to run.

The students continue completing their worksheets; talking becomes rather loud as discussion expands beyond close neighbors. While they are completing their worksheets, the solutions are displayed on the overhead projector.

Shortly before the end of the lesson, the teacher asks the students to pay attention to the distillation demonstration. He wants to prove that the distillation product actually is alcohol, so he tries to burn the distilled liquid. At first nothing at all burns. He heats it with the Bunsen burner and succeeds in igniting it. Once again, he shows the final results of the chromatography demonstration and gives some concluding explanations. The students are still attentive and ask further questions (e.g., "Which topics do we have to know by heart for the next test?"). Although the bell has already indicated the end of the lesson, they stay quiet and participate for a few more minutes. Then they leave the room, but two or three students stay around the demonstration desk continuing to talk with the teacher.

UNITED STATES:
Case Study

FOREWORD

The United States does not have a national curriculum in any subject. General curricular guidelines, teacher training and certification requirements, and length of the school year are some of the many aspects of education that are determined at the state level (Valverde, 1995). Although local school districts enjoy a great deal of autonomy and independence in curricular and instructional decisions, there is a considerable degree of consistency among systems both within as well as across states. Despite the decentralization of curriculum decision-making, the US is in the midst of considerable discussion and debate concerning what are the appropriate topics to be included in the mathematics and science curricula particularly at the primary and lower secondary ("middle") school level (Blank & Pechman, 1995). This debate is spearheaded by science and mathematics reformers, who are calling for a more conceptually driven approach to the teaching of these subjects. Many states and local school districts have begun efforts to redefine, revise, and refocus both the curriculum and teaching of mathematics and science. In addition, many states have initiated major reform efforts directed at the way local schools are organized and funded. The cases reported below may be fairly typical of what is happening at the primary and lower secondary school level as teachers make marginal changes in what is otherwise fairly traditional instruction.

MATHEMATICS, POPULATION 1

Mrs. Hunt is a veteran teacher, having worked at the primary school level for 21 years. For the past three years she has worked in a combination fourth/fifth grade classroom. Her district is one of the largest in the state with a population of over 100,000 students. While the district is economically and racially diverse, her school is one of the more advantaged, being located in a predominantly upper-middle class neighborhood.

The room is arranged in a traditional manner. Desks are placed end to end to form four rows, all facing forward. The teacher's desk is in the upper right hand corner of the room, on the opposite side from the door leading to the out-side. The class is racially diverse, consisting of two Hispanic children, seven Asian children, and eleven Caucasian children – six boys and fourteen girls in all. A parent volunteer is near the middle of the far wall, across from where the observer sat.

> *Ethnic diversity is the rule in urban districts' classrooms in – even in relatively privileged schools like this one. It is not unusual in such priv-ileged schools to find parent volunteers who are willing to assist teach-ers by correcting papers, running errands, and the like.*

The teacher previewed the days activities. "How many need to work on cut-ting out fraction pieces?" she asked. She suggested that the students complete this task even though she was about to call on students to present their papers describing what they did over the break. "Ordinarily we wouldn't do this while people are presenting." She told those engaged in readying this mater-ial not to worry about coloring the pieces.

While students were reporting on their activities during the recent break period, the teacher listened intently, occasionally interjecting a remark of two. After some further talk with the class, she asked students about the plants they were supposed to be growing for an upcoming fair. "So everybody has some plants growing?" Aaron shook his head. "What happened?" the teacher asked. "Something ate my plant." Another child chimed in that, even though one of his plants was watered and was bug-free, "it, like, all shriveled up." Other children suggested what might have gone wrong, the consensus being that it might not have gotten enough sunlight.

The teacher then discussed the students' Father's Day projects. Children were doing needlework. This involved pasting a carefully numbered, light sheet of paper to one that was thicker and colored; children were to follow the pattern laid out on the lighter sheet in their stitching. The reverse side of the colored paper would, if they did what was expected, yield an ornately designed "DAD." As they worked on their math seatwork, they were to obtain their needles from the parent volunteer. She explained how they were to thread their needles, taking care to leave enough of the thread on the other side of the needle so as not to slip through the eye of the needle.

As the above suggests, teachers in primary classrooms in the US devote a fair amount of time to "hands on" activities, either of the "plant growing" variety in science or of the "arts and crafts" variety (i.e., the Father's Day needlework). There are various interpretations of why this is the case. Many scholars attribute it to the fact that primary teachers typically do not major in academic subjects--or, if they do, those subjects tend to be "softer" subjects like psychology or English language arts. The lack of a strong subject matter emphasis in their professional training may incline US teachers toward more of a "child-centered" as opposed to "subject-centered" pedagogy. Child-centeredness, in turn, often orients primary school teachers in the US toward approaches that attempt to build on the child's interest as an entree into more substantive, content-related instruction.

At this point, Mrs. Hunt begins the math lesson. She tells the students that, whether or not they have finished cutting out their fraction pieces, they are going to "do some fractions." She starts with a little review. She pulls down the overhead screen that was position above the blackboard at the front of the class. Standing beside the projector, she asks some questions to the class. She asks if they could tell her the difference between a fraction and a whole number. "It's cut up," one student volunteers. Others offer "...like a quarter, a half." "It's part of a whole," the teacher suggests, repeating this definition once more to make sure it was heard.

This type of question-answer-response exchange is fairly typical of US primary mathematics lessons. The teacher elicits a short, quick response from students, acknowledges its correctness or incorrectness, and than supplies what is regarded as the ideal response. The purpose behind this kind of exchange may be as much related to the need to maintain students' attention as it is to ascertain what they know or don't know about a particular aspect of the lesson.

The teacher points out that, even though they usually "work in circles" when doing fractions, you can "also take sets of things and divide them up. Let's do this together." Turning on the overhead projector, the teacher points to a worksheet that had been converted into a transparency. "What does one-eighth mean?...The whole is divided into..." "Into eight parts," several students say. "Then there is one part of eight shapes," she adds.

Next she points to a circle in the upper left hand corner of the worksheet that had been divided into thirds. "How many parts?" "Three." "This is one third..." she says, using her marker to write a one in the top-most section. "This is one third...this is one third."

Mrs. Hunt works through several more figures. She refers to one long rectangle that had been divided into twelve parts as "sheet cake," reminding them of a dessert they had had at a recent birthday celebration. "Imagine there are twelve people coming to a party," she says, elaborating on a "real life" situation. She then presented another example. She and her husband recently put up an arbor at their house; she had designed the arbor to be ten by eleven. Then she discovered that lumber came in twelve foot pieces. "I redesigned it so we could use a standard piece of lumber." This was supposed to be an example of using fractions to solve a problem but it was unclear whether or not students realized this.

US lessons at the primary level tend to be based on what is in the textbook. Only minimal suggestions are provided teachers in their special editions about how to proceed in a lesson. The teachers' edition of the textbook usually contains basic information about the lesson – such as the lesson objective, hints for how to get started on the lesson, a few sentences about how to proceed in teaching the lesson, and suggestions for possible follow up activities. This material is presented in the margins in the teacher's edition, around a reproduction of the material contained in the student text.

The fact that lessons are only minimally scripted, combined with the likelihood that teachers at the primary level are not specialists in the subjects they teach, leaves a lot of room for improvisation of the sort described above. The pedagogical value of such improvisation can greatly vary and is often strongly dependent upon the teacher's subject matter background.

The teacher continues working through examples on the overhead. She shades in the top third of a square-shaped figure: "How many parts?" "Three." The teacher writes a small "3" in the denominator position just below the figure. "How many shaded parts?" "One." "How many are not shaded?" She constructs a number sentence: "1/3 + 2/3." "If we add one-third and two-thirds, how many do we have?" "Three-thirds," one student offers.

The teacher works through four more examples on the overhead. "Okay, you've got that down. You are so good," she says. "What we are going to do next..." she begins, as she passes out a worksheet.

The worksheet has the heading "Prove It!" The directions state:

> First read the problem. Find the cutouts needed to solve the problem [i.e., the fraction pieces cut out earlier]. Lay the cutouts on your desk and use them to find the answer. Write your answer on the line. Fill in and color the diagrams to illustrate your answer.

Each of the eight problems consist of two blank circles. Above each pair of circles is a question. The first is an example: "How many eighths equal 1/2?" Under the left circle is the word "halves;" under the right circle is the word, "eighths." The teacher tells students they will need their pencils.

"Okay," she says returning to her seat by the overhead, "I'll do the first one with you. I wish I had see through pieces. You'll have to trust me," she says, joking about the fact that it is hard to see some of the pieces once they were assembled into a circle on the overhead projector. She asks a student to read the first problem. She then models the approach the students are to use, laying out two halves of a circle to form one whole on one side, eight eighths to form a whole on the other side. The second circle, given the smallness of the pieces, is a bit askew. The teacher picks up the half circle and places it over one half of the circle that had been constructed out of the eight pieces. This is hard to see. Using a different approach, she reaches over and removes half of the second circle.

Many teachers at the upper primary level in the US are just beginning to experiment with the use of manipulatives to teach topics like fractions. Previously, concrete experiences of the sort provided by manipulatives was regarded as less important for students in the fourth, fifth, or sixth grade. This assumption has been challenged by leaders of the mathematics reform movement currently underway in the US.

A student starts to suggest another approach. "Just a minute," the teacher admonishes. "Do it this way first." "Four eighths," another student replies answering her earlier question. The teacher explains her approach. She works through several more examples on the overhead then tells students to begin working on the worksheet. One boy shades only two-sixths of a circle which he had subdivided into six parts in response to a question. The answer to this problem, which they went over with the overhead, was four sixths. He explains how he had solved that problem to the girl sitting next to him.

Teachers in the US at the primary level often encourage this sort of cooperative learning on the part of students.

After a few minutes, the teacher interrupts with a suggestion for problem number three. "Remember, you're subtracting," she says, apparently in reference to her preferred strategy of replacing the required number of larger pieces with pieces from the target circle. The latter can then be counted to yield the correct response. She continues, "What might be an easier way?" She suggests that, instead of taking away three-quarters in this particular problem, they need only remove one-quarter. The smaller piece could then be laid on the sixteenths' pieces. "You're working too hard," she says. The students appear confused at this point.

Sensing the difficulty they were in, the teacher once again turns on the overhead projector. "Everybody, do this. How many one-fourths are there in a circle?" "Three," a student says, apparently referring to the problem ("How many sixteenths in 3/4?").

"No. First set it up in a circle," the teacher replies. She takes four pieces and constructs a circle on the overhead. "There's four fourths," she points out. "How many sixteenths?" Here she begins to construct a circle out of the sixteenth pieces available to her. Because the pieces are so small she has difficulty with this.

Eventually, the teacher manages to construct a whole out of the sixteenth pieces. "I have two circles, one of fourths and one of sixteenths. How many sixteenths equal three fourths?" She then removes three fourths of the circle on the left. She points out that this leaves one fourth. "What am I going to do with the one fourth?" she asks. "Where am I going to put it?" "On the sixteenths?" guesses one student. "Right." She makes this move – which is hard to see on the overhead.

"You can't see very well on the overhead," she notes. "I probably shouldn't do it with sixteenths. See, basically what I've done here...It's just a matter of counting the sixteenths." The teacher demonstrates how the quarter piece accounted for four of the smaller sixteenth pieces.   She then writes "3/4 = 12/16" on the overhead and explains, referring back to the part of the circle not covered by the one-fourth piece, "You count the part that's not covered."

"Let's go ahead with number four," the teacher says. "Trevor, tell us what you did." This problem reads: "How many thirds equal 8/12?"

"Okay. You first drew twelfths, then thirds, then you colored in eight of the twelfths..." the teacher repeats what the student says.   Trevor continues: "I put the twelfths together and the thirds." He describes how he used the fraction pieces.

"Why am I hearing talk?" the teacher interjects, silencing the class.

"I put together the twelfths; I put together the thirds; I took away eight twelfths. I then laid the thirds on top." After calling for one more example, the teacher decides to draw the discussion to a close. "The main thing," she says, after indicating that they would return to the problems later, "is don't lose any little pieces. Put them in the envelopes."

SCIENCE, POPULATION 2

Steve Shipley is a veteran teacher who has taught science in this school district for over twenty years ever since he graduated from a state university. With a dual major in biology and education, he is one of three science teachers at Township Middle School who either majored or minored in a science as undergraduates. The other two science teachers majored in primary education but have subsequently focused their teaching and professional development activities on teaching science to students in grades seven and eight. One of these two, who is currently the lead science teacher (i.e., department head) for the school, also has a Masters' degree in science education from another state university.

Mr. Shipley's classroom is one of three classrooms that have regularly spaced experiment stations equipped with lab tables; water, air and gas outlets; and a sink. Students sit on lab stools at these tables. While this arrangement is great for doing experiments, students often find it uncomfortable and distracting during lectures, class discussions, and exams. Mr. Shipley expresses great appreciation for his district's commitment to science education because he has always either had access to or been able to purchase any materials he has needed for lab experiments and demonstrations. He contrasts this to his sister's situation on the other side of the state where her district has a high concentration of students from migrant agricultural families and funds for such resources are always tight.   In fact, she often buys supplies for labs and demonstrations with her own personal funds.

Mr. Shipley stands in the hall outside his classroom and greets each of his students as they enter. Although this is the second class period of the day, it is Mr. Shipley's first class since his planning period is first.

> *In most lower secondary schools, teachers usually have their own classrooms and students move from one classroom to another during the break between class periods which is typically 5 minutes.   There are between six and eight instructional class periods in a normal school day with each period lasting 50 or 55 minutes.   Students are assigned lockers, small closets or compartments in which books, supplies, and other personal belongings may be stored.   Lockers are typically located in the school's hallways.*

Due to the brief amount of time between classes, students do not have time to visit their lockers and must carry all their books and necessary supplies with them. As students enter the room, they place their large backpacks on the shelves at the rear of the room and take a notebook and pencil with them to their assigned seats around the lab tables. Since he has hall duty this week, Mr. Shipley remains in the hall for another 5 minutes after the bell that signals the beginning of the next class period has rung in order to make sure all students are in classrooms and no one is left wandering the halls.

> *Hall duty is one of many different types of general supervisory responsibilities teachers are sometimes required to provide in the school.   Teachers who have hall duty check the school's hallways to ensure that all students are in class and that there are no unauthorized outsiders wandering the school.   Typically this type of duty is assigned to teachers on a rotating basis for a period of one week.   Teachers may also provide this type of supervision in the school's cafeteria and at special scheduled events such as sports meets or fine arts performances.   Teachers sometimes must forgo their planning or lunch time in order to carry out these supervisory responsibilities.   In many districts, teachers' labor unions have succeeded in eliminating this type of supervision from teachers' responsibilities.*

Once his hall duty responsibility has been satisfied, Mr. Shipley enters his class and announces, "Okay, its time to begin. Everyone please take your seats and take out the sheets on the size of molecules and the one on atoms and isotopes I gave you yesterday as homework for today. We'll go over the size of molecules sheet first."

Mr. Shipley then takes a moment to check students' attendance. Twenty-one students are present, and five students are absent. Three students are absent today because they are on a special field trip as part of the ESL (English as a Second Language) program. This program offers special educational support to those students who come from homes where English is not the first language. Mr. Shipley quickly checks the "Excused Absence" list that the school office has distributed to make sure the other two students are listed. Parents must call the school office before the beginning of the first period to have their students name put on the "Excused Absence" list if a student will not be able to attend school that day due to an illness or some other legitimate reason. Otherwise, students are considered to have been absent for illegitimate reasons and face disciplinary consequences.

Teachers are typically required to maintain records of students' class attendance and the number of times a student comes to class after the class has begun. Teachers must report student absences and late arrivals to the school's administrative office. Schools usually have disciplinary consequences for students who exceed a certain number of late arrivals or unexcused absences. Other typical administrative duties include maintaining an inventory of instructional and laboratory resources and materials and reporting to the proper authorities any suspicions of mistreatment or neglect of a student by parents or guardians.

Most of today's lesson centers on the review and correction of two worksheets Mr. Shipley handed out the day before which were to be completed as homework. The worksheets are two exercises that were included in the supplementary materials kit that accompanied the Teacher's Edition of the textbook.

Textbook publishers frequently produce a supplementary materials kit that accompanies the Teacher's Edition of the student text. The Teacher's Edition often contains answers to exercises found in the student text as well as suggestions for teaching particular topics and lesson activities.

An accompanying supplementary materials kit may contain overheads, worksheets to be reproduced for students' use, supplies for a special projects or activities, posters, and other items for display or demonstration.

For over ten years, Mr. Shipley taught general science to students in grades seven and eight focusing on slightly different topics for each grade. However, since the local school district revised the curriculum five years ago, he has had primary responsibility to teach physical science to the eighth grade students. (Seventh grade students now study life science while sixth grade students continue to follow a diverse curriculum encompassing earth, physical, and life sciences.)

While he uses the same textbook and lesson plans with all his classes, he finds that he must occasionally simplify his expectations for the two slow classes and increase his expectations for students in the advanced class. Second period is one of Mr. Shipley's three average classes.

Starting in lower secondary school (i.e., grades 6, 7, and 8), the tracking of students into high, middle, and low groups is more the norm than the exception in the so-called academic subjects (i.e., English, math, science, and social studies). Despite wide-spread condemnation of this practice by educational reformers, tracking remains a common practice in US schools.

Some students look at the board at the front of the room where Mr. Shipley has written a list of "Coming Attractions", a list of topics and activities which will be covered in coming weeks. There is also a list entitled "This Week" that lists the topics and assignments for each day of the current week.

Having quickly completed his check of students' attendance, Mr. Shipley asks the students how many were able to complete the two homework sheets. A few students raise their hands. He announces that they will continue their unit on atoms and molecules by going over the homework sheets together. Mr. Shipley has overhead copies of the students' homework sheets. He places one of the homework sheets he handed out yesterday on the overhead, turns the overhead projector on, and turns the classroom lights down. This sheet has 10 problems on it concerning the size of atoms and molecules.

The first problem reads:

> An elite typewriter period contains one trillion molecules
> of ink. If the volume of each period is 0.00000006 cm^3,
> what would be the volume of each ink molecule?"

Mr. Shipley asks students to tell how they would do this problem and why they would set it up the way they suggest. A few students offer their ideas. He then asks students to consider the relationship between the size of one molecule and the volume of the period. He asks them to consider the information given and to estimate the size of a molecule to the closest power of 10, e.g., 10^{-6} or 10^{-30}. Students offer various ideas and discuss the reasonableness of different estimates.

Having come to a consensus response to the first question, Mr. Shipley moves on to the second question:

> The average human inhales 1,000 cm^3 of air with each
> breath. Each 1,000 cm^3 of air contains 200 cm^3 of oxygen
> (O_2). If each molecule of O_2 is 0.00000000000000000004
> cm^3, how many molecules of O_2 are inhaled with each
> breath?

Mr. Shipley again asks students how they would approach the problem and why they would use that approach. He then asks them to estimate a reasonable answer to the closest power of 10.

The loudspeaker in the room interrupts the lesson with an announcement from the office. Due to the State Competency Test that the seventh graders are taking today, there is a change in the lunch schedule. All seventh graders will eat lunch during lunch-period A, before the beginning of Fourth Period; all sixth graders will eat during lunch-period B, the twenty-five minute period in the middle of Fourth Period; and all eighth graders will eat during lunch-period C, the last twenty-five minutes of Fourth Period. (Unlike every other period, Fourth Period is seventy-five minutes long to accommodate a twenty-five minute lunch break.) Mr. Shipley and the students take this interruption from the loudspeaker in stride, pausing their lesson discussion just long enough to attend to the announcement.

Schools frequently lengthen the class period around noon to accommodate scheduling lunch times for different student groups. Schools may have two, three, or four different lunch times depending on the number of students in the school and the capacity of the available cafeteria facilities.

Mr. Shipley continues to review three more problems from the first homework sheet students were to have completed for today. Once he is convinced that students understand what needs to be done and that they are capable of doing this type of problem, he announces that they will move on to the second of yesterday's homework sheets. This one is about atomic number and is entitled, "Isotopes or Different Elements?" He turns the lights back on and begins by reviewing the definitions of an isotope and the parts of an atom: protons, neutrons, and electrons. He has a plastic model of an atomic nucleus and uses it to illustrate the concepts of atomic mass, atomic number, protons, and neutrons. He then asks students to look at their homework sheet and quickly works through it with the whole class. The homework handout contains descriptions of seven pairs of elements. For each pair, students must decide whether the pair are different elements or different isotopes of the same element. For example, the fifth pair reads:

Element T has an atomic number of 20 and an atomic mass of 40; Element Z has an atomic number of 20 and an atomic mass of 41.

When they finish reviewing the second homework sheet, Mr. Shipley goes over the plans for each class period for the week. According to the schedule he has put on the blackboard, they should complete their introduction to the structure of matter by the end of the week. Mr. Shipley reminds the class that next week is National Chemist's Week and he expects a professional chemist to visit all the science classes either Tuesday or Wednesday to talk to students about chemistry and careers related to chemistry. Last year, Mr. Shipley tells the class, the chemist who came was a woman from a nearby well-known chemical company who explained how plastic food wrap was made and demonstrated the manufacture of a synthetic polymer. The bell rings signaling the end of second period. Students begin to gather their materials and exit the classroom. Mr. Shipley stands by the door and hands each student a homework handout for tomorrow's class concerning probability and the location of electrons.

APPENDIX A

THE TIMSS CURRICULUM FRAMEWORKS

The curriculum frameworks were developed for use in the Third International Mathematics and Science Study (TIMSS). This extensive study conducted in over 40 countries involved three main components: an analysis of curriculum documents, a set of school, teacher, and student questionnaires, and student achievement testing. The scope of this effort required that all TIMSS components be linked by a common category framework and descriptive language. Whether classifying a test item, characterizing part of a curriculum document, or linking a questionnaire item to other TIMSS parts, any description had to use common terms, categories, and standardized procedures so numerical codes could be assigned for entry into the appropriate database.

The two curriculum frameworks— one for mathematics and one for the sciences — represent a multi-category, multi-aspect specification of these two subjects that provide a common language system for TIMSS. Both include multiple categories for each of three aspects of subject matter— *content* (specific subject matter topics), *performance expectation* (what students are expected to do with particular topics), and *perspective* (the attitudes or perspectives encouraged or promoted with respect to the subject matter and its place among the disciplines and in the everyday world).

The frameworks were designed to accommodate the use of multiple categories from each aspect. Any curricular element such as a curriculum guide, textbook segment, or test item can be classified in as many framework categories as needed from all three aspects to capture its richness. In this way each element can have a unique "signature" — a set of content, performance expectation, and perspective categories that characterize it. This flexible system allows simple or complex signatures as needed. This differs from traditional "grid" classification systems that generate unique categorizations that combine a single element from two or more dimensions.

The frameworks were designed to be suitable for all participating countries and educational systems. Representing the interests of many countries, the frameworks were designed cross nationally and passed through several iterations. The result is imperfect, but still is a step forward in cross-national comparisons of curricular documents.

Each aspect of the frameworks is organized hierarchically using nested sub-categories of increasing specificity. Within a given level, the arrangement of topics does not reflect a particular rational ordering of the content. Each framework aspect was meant to be encyclopedic in terms of covering all possibilities at some level of specificity. No claim is made that the "grain size" — the level of specificity for each aspect's categories — is the same throughout the frameworks. Some sub-categories are more inclusive and commonly used, others less so.

The mathematics framework contains 10 major *content* categories each with 2 to 17 sub-categories. Some sub-categories are divided still further resulting in a total of 44 different topics specified at the lowest possible level. The science framework specifies 8 major *content* categories further divided into 2 to 6 sub-categories. One major category, History of Science and Technology, has no sub-categories. The total number of science topics specified at the lowest level is 77. In each case, the level of detail and organization reflects a compromise between simplicity (fewer categories) and specificity (more categories). The hierarchical levels of increasing specificity allow a degree of flexibility in detail level and generalization.

The mathematics framework contains 5 major *performance expectation* categories each with 3 to 6 sub-categories and 5 *perspective* categories. One *perspective* category has two subdivisions. The science framework contains 5 major *performance expectation* categories each with 2 to 5 sub-categories and 6 *perspective* categories. Two of the *perspective* categories have two subdivisions.

What follows is an illustration of the hierarchical organization of the three framework aspects for both mathematics and science. Readers are referred to the *Curriculum Frameworks for Mathematics and Science* monograph listed in the references for a more detailed discussion or to the technical report of "Explanatory Notes" for the mathematics and science frameworks (McKnight, 1992; Britton, 1992).

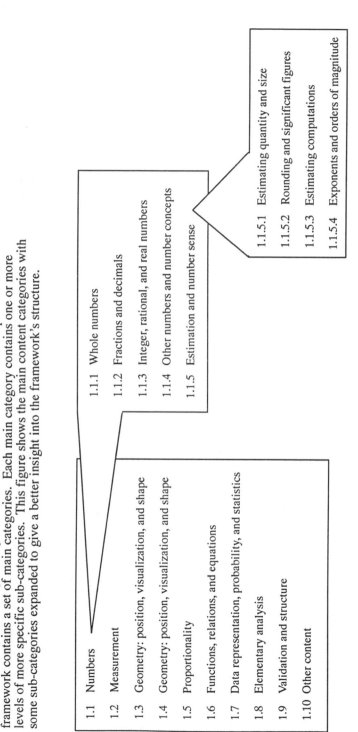

Figure A-1. Content Categories of the Mathematics Framework. Each aspect of the framework contains a set of main categories. Each main category contains one or more levels of more specific sub-categories. This figure shows the main content categories with some sub-categories expanded to give a better insight into the framework's structure.

1.1   Numbers

    1.1.1   Whole numbers

    1.1.2   Fractions and decimals

    1.1.3   Integer, rational, and real numbers

    1.1.4   Other numbers and number concepts

    1.1.5   Estimation and number sense

        1.1.5.1   Estimating quantity and size

        1.1.5.2   Rounding and significant figures

        1.1.5.3   Estimating computations

        1.1.5.4   Exponents and orders of magnitude

1.2   Measurement

1.3   Geometry: position, visualization, and shape

1.4   Geometry: position, visualization, and shape

1.5   Proportionality

1.6   Functions, relations, and equations

1.7   Data representation, probability, and statistics

1.8   Elementary analysis

1.9   Validation and structure

1.10 Other content

Figure A-2. Content Categories of the Science Framework. Simliar to Figure A-1, this figure shows the main science categories with some sub-categories expanded to give a better insight into the framework's structure.

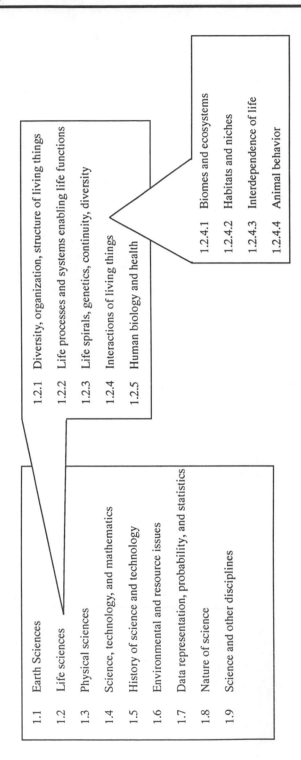

1.1	Earth Sciences
1.2	Life sciences
1.3	Physical sciences
1.4	Science, technology, and mathematics
1.5	History of science and technology
1.6	Environmental and resource issues
1.7	Data representation, probability, and statistics
1.8	Nature of science
1.9	Science and other disciplines

1.2.1	Diversity, organization, structure of living things
1.2.2	Life processes and systems enabling life functions
1.2.3	Life spirals, genetics, continuity, diversity
1.2.4	Interactions of living things
1.2.5	Human biology and health

1.2.4.1	Biomes and ecosystems
1.2.4.2	Habitats and niches
1.2.4.3	Interdependence of life
1.2.4.4	Animal behavior

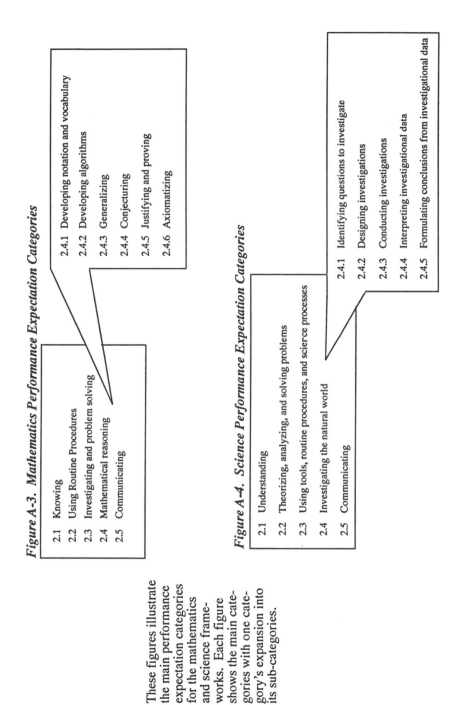

Figure A-3. Mathematics Performance Expectation Categories

2.1   Knowing

2.2   Using Routine Procedures

2.3   Investigating and problem solving

2.4   Mathematical reasoning

2.5   Communicating

2.4.1   Developing notation and vocabulary

2.4.2   Developing algorithms

2.4.3   Generalizing

2.4.4   Conjecturing

2.4.5   Justifying and proving

2.4.6   Axiomatizing

Figure A-4. Science Performance Expectation Categories

2.1   Understanding

2.2   Theorizing, analyzing, and solving problems

2.3   Using tools, routine procedures, and science processes

2.4   Investigating the natural world

2.5   Communicating

2.4.1   Identifying questions to investigate

2.4.2   Designing investigations

2.4.3   Conducting investigations

2.4.4   Interpreting investigational data

2.4.5   Formulating conclusions from investigational data

These figures illustrate the main performance expectation categories for the mathematics and science frameworks. Each figure shows the main categories with one category's expansion into its sub-categories.

Figure A-5. Mathematics Perspective Categories

3.1	Attitudes towards science, mathematics, and technology
3.2	Careers involving science, mathematics, and technology
3.3	Participation in science and mathematics by underrepresented groups
3.4	Science, mathematics, and technology to increase interest
3.5	Scientific and mathematical habits of mind

Figure A-6. Science Perspective Categories

3.1	Attitudes towards science, mathematics, and technology
3.2	Careers in science, mathematics, and technology
3.3	Participation in science and mathematics by underrepresented groups
3.4	Science, mathematics, and technology to increase interest
3.5	Safety in science performance
3.6	Scientific habits of mind

APPENDIX B

SMSO RESEARCH REPORT SERIES INDEX

#1	Compilation of Items Measuring Mathematics and Science Learning Opportunities and Classroom Processes from Large-Scale Educational Surveys, September 17, 1991
#2	Mathematics Curriculum Framework (McKnight and Swafford), July 1991(ICC322/NPC088)
#3	Science Curriculum Framework (Raizen and Britton), July 1991 (ICC307/NPC080)
#4	Topic Trace Mapping Instructions and Forms (McKnight), July 1991 (NPC080)
#5	Document Analysis Manual (Field Trial Version) (McKnight), July 1991 (NPC089)
#6	Modified Topic Trace Mapping Instructions and Forms (McKnight), August 1991 (NPC041)
#7	Mathematics Curriculum Framework (Explanatory Notes) (McKnight), September 1991 (ICC321/NPC087)
#8	Supplement to Document Analysis Manual (Field Trial Version) (McKnight), September 1991
#9	OTL Framework (Schmidt), Notes on SMSO Instructional Practices (Putnam), SMSO Teacher Knowledge and Beliefs (Prawat), Prepared in Advance for SMSO Planning Meeting in East Lansing, MI, September 1991
#10	Instructional Practices: Conceptual Matrix, September 21, 1991
#11	Conceptual Framework: OTL and Content Coverage Goals Notes, SMSO Planning Meeting, Washington, D.C., September 27/28, 1991
#12	Document Analysis Manual (Field Trial Version) (McKnight), October 1991 (ICC326/NPC089)
#13	Science Curriculum Framework as revised by the U.S. Steering Committee, October 1991
#14	Mathematics Curriculum Framework as revised by the U.S. Steering Committee, October 1991
#15	Teacher Beliefs, Draft 1
#16	Instructional Practices General Teacher Questionnaire, Draft 1
#17	Instructional Practices Lesson Specific Teacher Questionnaire, Draft 1
#18	Topic Coverage Questionnaire, Draft 1

#19	Teacher Beliefs Battery: Situated Questions on Teaching Selected Mathematics and Science Content, Draft 2 (Supercedes #15)
#20	Answer Sheet for OTL Instrument: Mathematics and Science, Populations 1 and 2, Draft 1
#21	Content Coverage Goals Questionnaire, Draft 2
#22	Instructional Practices General Teacher Questionnaire, Draft 2 (Supercedes #16)
#23	Instructional Practices Lesson Specific Teacher Questionnaire (Snapshot), Draft 2 (Supercedes #17)
#24	Topic Specific Questionnaire, Draft 2
#25.1.M	Instructional Practices General Teacher Questionnaire, POP 1: Mathematics, Draft 3 (Supercedes Reports No. #16 and #22), March 12, 1992
#25.1.S	Instructional Practices General Teacher Questionnaire, POP 1: Science, Draft 3 (Supercedes Reports #16 and #22), March 12, 1992
#25.2.M	Instructional Practices General Teacher Questionnaire, POP 2: Mathematics, Draft 3 (Supercedes Reports #16 and #22), March 12, 1992
#25.2.S	Instructional Practices General Teacher Questionnaire, POP 2: Science, Draft 3 (Supercedes Reports #16 and #22), March 12, 1992
#26	Document Analysis Manual (draft) (McKnight, Britton, Valverde, Schmidt), Supercedes Reports #5 and #12), March 16, 1992
#27	Mathematics Framework, Draft 3 (Supercedes Reports #2 and #14), March 3, 1992
#28	Science Framework, Draft 3 (Supercedes Reports #3 and #13), March 3, 1992
#29.1.S	Teacher Beliefs Battery, POP 1: Science, Draft 3 (Supercedes Reports #15 and #19), February 28, 1992
#29.1.M	Teacher Beliefs Battery, POP 1: Mathematics, Draft 3 (Supercedes Reports #15 and #19), February 28, 1992 (NPC068)
#29.2.M	Teacher Beliefs Battery, POP 2: Mathematics, Draft 3, (Supercedes Reports #15 and #19), February 28, 1992 (NPC069)
#29.2.S	Teacher Beliefs Battery, POP 2: Science, Draft 3, (Supercedes Reports #15 and #19), February 28, 1992 (NPC071)
#30	Topic Trace Manual (draft) (McKnight, Valverde, Schmidt), March 3, 1992
#31.1.M	Instructional Practices, Lesson Specific Teacher Questionnaire (Snapshot), POP 1: Mathematics, Draft 3 (Supercedes Reports #17 and #23), March 23, 1992
#32	Topic Specific, Draft 2 (Supercedes Report #24), March 12, 1992

#33.1.M	Instructional Practices Lesson Specific Teacher Questionnaire (Snapshot), POP 1: Mathematics, Draft 4, (Supercedes Reports #17, #23 and #31), April 1992 (NPC074)
#33.1.S	Instructional Practices Lesson Specific Teacher Questionnaire (Snapshot), POP 1: Science, Draft 4, (Supercedes Reports #17, #23, and #31), April 1992
#33.2.M	Instructional Practices Lesson Specific Teacher Questionnaire (Snapshot), POP 2: Mathematics, Draft 4, (Supercedes Reports #17, #23 and #31), April 1992
#33.2.S	Instructional Practices Lesson Specific Teacher Questionnaire (Snapshot), POP 2: Science, Draft 4, (Supercedes Reports #17, #23 and #31), April 1992
#34.1.M	Topic Specific Questionnaire, POP 1: Mathematics, Draft 3, (Supercedes Reports #24 and #32), April 1992
#34.1.S	Topic Specific Questionnaire, POP 1: Science, Draft 3, (Supercedes Reports #24 and #32), April 1992
#34.2.M	Topic Specific Questionnaire, POP 2: Mathematics, Draft 3, (Supercedes Reports #24 and #32), April 1992
#34.2.S	Topic Specific Questionnaire, POP 2: Science, Draft 3, (Supercedes Reports #24 and #32), April 1992
#35.1.M	Opportunity to Learn Teacher Questionnaire, POP 1: Mathematics, Draft 2, (Supercedes Report #20), April 1992 (ICC376/NPC110)
#35.1.S	Opportunity to Learn Teacher Questionnaire, POP 1: Science, Draft 2, (Supercedes Report #20), April 1992
#35.2.M	Opportunity to Learn Teacher Questionnaire, POP 2: Mathematics, Draft 2, (Supercedes Report #20), April 1992
#35.2.S	Opportunity to Learn Teacher Questionnaire, POP 2: Science, Draft 2, (Supercedes Report #20), April 1992 (ICC379/NPC113)
#36.1.M	Teacher Beliefs and Attitudes Questionnaire, POP 1: Mathematics, Draft 4, (Supercedes Reports #15, #19 and #29), April 1992
#36.1.S	Teacher Beliefs and Attitudes Questionnaire, POP 1: Science, Draft 4, (Supercedes Reports #15, #19 and #29), April 1992
#36.2.M	Teacher Beliefs and Attitudes Questionnaire, POP 2: Mathematics, Draft 4, (Supercedes Reports #15, #19 and #29), April 1992
#36.2.S	Teacher Beliefs and Attitudes Questionnaire, POP 2: Science, Draft 4, (Supercedes Reports #15, #19 and #29), April 1992
#37	Science Curriculum Framework, Draft 4, (Supercedes Reports #3, #13 and #28), May 1992
#38	Mathematics Curriculum Framework, Draft 4, (Supercedes Reports #2, #14 and #27), May 1992 (ICC307NPC080)
#39	International Planning Meeting - Minutes, February 5-8, 1992, East Lansing, Michigan

#40	Explanatory Notes for the TIMSS Science Framework, May 1992
#41	Explanatory Notes for the TIMSS Mathematics Framework, May 1992 (NPC140)
#42	Document Analysis Manual, Draft 4, (Supercedes Reports #5, #12 and #26), May 1992 (ICS326/NPC089)
#43	In-depth Topic Trace Mapping, Draft 4, (Supercedes Reports #4, #6 and #30), May 1992(ICC406/NPC139)
#44	Training Manual, Document Analysis, In-Depth Topic Trace Mapping, Regional Training Meetings, (McKnight and Britton), June 1992
#45.1	Instructional Practices and Opportunity to Learn, POP 1, Draft 1, August 1992
#45.2.M	Instructional Practices and Opportunity to Learn, POP 2: Mathematics, Draft 1, July 1992
#45.2.S	Instructional Practices and Opportunity to Learn, POP 2: Sciences, Draft 1, July 1992
#46.1	Teacher Pedagogical Beliefs and Opportunity to Learn, POP 1, Draft 1, August 1992
#46.2.M	Teacher Pedagogical Beliefs and Opportunity to Learn, POP 2: Mathematics, Draft 1, July 1992
#46.2.S	Teacher Pedagogical Beliefs and Opportunity to Learn, POP 2: Sciences, Draft 1, July 1992
#47.1	Teacher Background and Classroom Practices, POP 1, Draft 1, August 1992
#47.2.M	Teacher Background and Classroom Practices, POP 2: Mathematics, Draft 1, July 1992
#47.2.S	Teacher Background and Classroom Practices, POP 2: Science, Draft 1, July 1992
#47.3B.P	Teacher Background and Classroom Practices, POP 3B: Physics, Draft 1, August 1992
#47.3B.M	Teacher Background and Classroom Practices, POP 3B: Mathematics, Draft 1, August 1992
#48.1.M/S	Instructional Practices, POP 1: Mathematics and Science, (Supercedes all prior Instructional Practices Questionnaires), August 1992
#48.2.M	Instructional Practices General, POP 2: Mathematics, Draft 4 (Supercedes Reports #16, #22 and #25.2.M), August 1992
#48.2.S	Instructional Practices General, POP 2: Science, Draft 4, (Supercedes Reports #16, #22 and #25.2.S), August 1992
#48.3B.M	Instructional Practices, POP 3B : Mathematics (Supercedes all prior Instructional Practices Questionnaires), August 1992
#48.3B.P	Instructional Practices, POP 3B: Physics (Supercedes all prior Instructional Practices Questionnaires), August 1992

#49.1.M/S	Opportunity to Learn, POP 1: Mathematics and Science, Draft 3, (Supercedes Reports #20, #35.1.M and #35.1.S), August 1992
#49.2.M	Opportunity to Learn, POP 2: Mathematics, Draft 3, (Supercedes Reports #30 and 35.2.M), August 1992
#49.2.S	Opportunity to Learn, POP 2: Science, Draft 3, (Supercedes Reports #30 and 35.2.S), August 1992
#50.1	Teacher Beliefs and Attitudes, POP 1, Draft 1, August 1992
#50.2.M	Teacher Beliefs and Attitudes, POP 2: Mathematics, Draft 5 (Supercedes Reports #15, #19, #29 .2.M and #36.2.M)
#50.2.S	Teacher Beliefs and Attitudes, POP 2: Science, Draft 5 (Supercedes Reports #15, #19, #29 .2.S and #36.2.S)
#50.3B.M	Teacher Beliefs and Attitudes, POP 3B: Mathematics, Draft 1, August 1992
#50.3B.P	Teacher Beliefs and Attitudes, POP 3B: Physics, Draft 1, August 1992
#51.1.2	Teacher Background, POPULATIONS 1 and 2, Draft 1, August 1992
#51.3B.P	Teacher Background, POP 3B: Physics, Draft 1, August 1992
#51.3B.M	Teacher Background, POP 3B: Mathematics, Draft 1, August 1992
#52	Content Coverage Goals: POP S 1 and 2: Mathematics and Science, August 1992 (NPC077,NPC078,NPC079)
#53	TIMSS Curriculum Analysis Training Manual Supplement, November 1992
#54	International Planning Meeting Minutes, Tokyo, June 24-26, 1992
#55	Teacher Context Questionnaires and Background Notes, November 1992
#56	TIMSS: Concepts, Measurements and Analyses, February 1993
#56	TIMSS: Concepts, Measurements and Analyses, (Abbreviated Version)
#56	TIMSS : Concepts, Measurements and Analyses, Revision of Appendix C
#57	TIMSS Curriculum Analysis: A Content Analytic Approach, March 1993
#58	TIMSS Educational Opportunity Model: Detailed Instrumentation and Indices Development, September 1993 (ICC713/NPC276)
#59	A Description of the TIMSS' Achievement Test Content Design - Test Blueprints (ICC797/NRC357)
#60	Minutes SMSO - International Meeting, Madrid, Spain, December 9-11, 1992
#61	Minutes SMSO - Paris Meeting, France, January 10-14, 1994
#62..XXX	Curriculum Analysis Data Review, July 1, 1994

SMSO CURRICULUM ANALYSIS
TECHNICAL REPORT SERIES INDEX

REFERENCES

References preceded by an asterisk (*) provide important background for the concepts and country case studies presented in this volume but have not been included in any specific citation within the text.

Anderson, L. W., Ryan, D. W., & Shapiro, B. J. (Eds.). (1989). *The IEA classroom environment study*. (Vol. 4). Oxford, England: Pergamon Press.

*Barrier, E., & Robin, D. (1987). L'enseignment des mathématiques dans le contexte international. Contribution à la réflexion (The teaching of mathematics in the international context. Contribution with critique). *Revue Française de Pédagogie, 80*, 5-15.

Beck, E., Guldimann, T., & Zutavern, M. (1991). Eigenständig lernende Schülerinnen und Schüler (Autonomous learners). *Zeitschrift für Pädagogik, 37*, 735-767.

*Birkemo, A., Grøterud, M., Hauge, T. E., Knutsen, A. E., & Nilsen, B. S. (1994). *Læringskvalitet i skolen (Learning in schools)* (International School Effectiveness Research Project 1). Oslo, Norway: University of Oslo.

Bjørndal, I. R. (1995). Norway: Sytem of education. In T. Husen & T. N. Postlethwaite (Eds.), The International Encyclopedia of Education, (pp. 4185-92). Oxford, England: Pergamon.

Blank, R. K. & Pechman, E. M. (1995). State Curriculum Frameworks in Mathematics and Science: How are They Changing Across the States? Washington, D.C: Council of Chief State School Officers.

Bloom, B. S., Engelhart, M. S., Furst, E. J., Hill, W. H., & Krathwohl, D. R. (1956). *Taxonomy of Educational Objectives: Handbook I: Cognitive Domain*. New York: Longman, Green.

Britton, E. D. (1992). *Explanatory notes for the science framework* (Survey of Mathematics and Science Opportunities Research Report Series #40): Michigan State University.

Bruner, J. S. (1966). *Toward a Theory of Instruction*. Cambridge, Mass.: Harvard University Press.

Burstein, L. (Ed.). (1993). *The IEA study of mathematics III: student growth and classroom processes*. (Vol. 3). Oxford, England: Pergamon Press.

Burstein, L., Oakes, J., & Guiton, G. (1992). Education indicators. In M. Alkin (Ed.), *Encyclopedia of educational research*, (6th ed., pp. 409-418). New York: Macmillan.

Carraher, T. N., Carraher, D. W., & Schlieman, A. D. (1987). Written and oral math. *Journal for Research in Mathematics Education, 18*, 83-97.

Clark, C. M., & Peterson, P. L. (1986). Teachers' thought processes. In M. C. Wittrock (Ed.), *Handbook of Research on Teaching*, (3rd ed., pp. 255-296). New York: Macmillan.

Clune, W. H. (1993). The best path to systemic educational policy. *Educational Evaluation and Policy Analysis, 15*, 233-254.

Coll, C. (1987). *Psicología y curriculum*. Barcelona: Laia.

Contreras, J. (1990). Enseñanza, curriculum y profesorado. Madrid: Akal.

Csikszentmihalyi, M. (1990). *Flow: The psychology of optimal experience*. New York: Harper and Row.

D'Ambrosio, U. (1985,). *Socio-cultural bases for mathematics education*. Paper presented at the 5th International Congress of Mathematics.

Deci, E. L., & Ryan, R. M. (1993). Die Selbstbestimmungstheorie der Motivation und ihre Bedeutung für die Pädagogik (Self-determination theory and its significance for pedagogy). *Zeitschrift für Pädagogik, 39*, 223-237.

Doyle, W. (1986). Classroom organization and management. In M. C. Wittrock (Ed.), *Handbook of Research on Teaching*, (Third ed., pp. 392-431). New York: Macmillan.

Edelstein, W. (1992). Development as the aim of education–revisited. In F. K. Oser, A. Dick, & J. Patry (Eds.), *Effective and Responsible Teaching*. San Francisco: Jossey-Bass.

EDK. (1995). *Neue Unterrichts- und Organisationsformen (New approaches of teaching and organizing instruction)* . Bern: Schweizerische Konferenz der kantonalen Erziehungsdirektoren (EDK).

Elley, W. B. (Ed.). (1992). *How in the world do students read? : IEA study of reading literacy*. Hamburg: The International Association for the Evaluation of Educational Achievement.

Elley, W. B. (Ed.). (1994). *The IEA study of reading literacy: achievement and instruction in thirty-two school systems*. Oxford, England/Tarrytown, NY: Pergamon Press.

Farnham-Diggory, S. (1994). Paradigms of knowledge and instruction. *Review of Educational Research, 64*, 463-477.

Federal Statistical Office. (1991). *The Swiss Educational Mosaic*. Bern: Federal Statistical Office.

Flanders, J. R. (1987). How much of the content in mathematics textbooks is new? *Arithmetic Teacher, 35*, 18-23.

*Fossland, T. N. (1994). *Konstruktivisme i klasserommet: teoretiske betrakt-ninger og en empirisk undersøkelse av naturfagundervisning (Constructivism in the classroom: theoretical considerations and an empirical study of science teaching)*. Unpublished Masters Thesis, University of Oslo, Oslo.

Gimeno, S. J., & Pérez Gómez, A. (1992). *Comprender y transformar la enseñanza*. Madrid: Morata.

Goetz, J. P., & LeCompte, M. D. (1984). *Ethnography and qualitative design in educational research*. Orlando, Fl: Academic Press.

Gretler, A. (1995). Switzerland: System of education. In T. Husen & T. N. Postlethwaite (Eds.), The International Encyclopedia of Education, (pp. 5873-83). Oxford, England: Pergamon.

Hernández, F., & Sancho, J. M. (1993). *Para enseñar no basta con saber la asígnatura*. Barcelona: Paidós.

*Jorde, D. (1992). Har vi klart å integrere vekk naturfag fra barnetrinnet? (Have we managed to integrate science out of primary educa-tion?). *Norsk Skoleblad, 26*, 26-28.

*Jorde, D., & Lea, A. (1995). Sharing science: Primary science for both teach-ers and pupils. In L. Parker & L. Rennie (Eds.), *Shortening the Shadows*, (pp. 155-166). Dordrecht: Kluwer.

*Kanaya, T. (1995). Japan: System of education. In T. Husen & T. N. Postlethwaite (Eds.), The International Encyclopedia of Education, (pp. 3078-86). Oxford, England: Pergamon.

Klippert, H. (1993). Förderung von Selbständigkeit und Selbststeuerung (Encouragement of self-independence and self-regulation). *Neue Sammlung, 3*, 437-454.

*Knutsen, A. E. (1993). *Lokale læreplaner som implementeringsstrategi i naturfag (Regional curriculum planning as a stragegy for imple-menting science teaching)*. Unpublished Masters Thesis, University of Oslo.

*Krohg–Sørensen, R. (1992). *Læreres tenkning og handling i naturfagun-dervisning på ungdomstrinnet: et kasusstudium (Teacher think-ing and handling in science teaching at the junior high school level: a case study)*. Unpublished Masters Thesis, University of Oslo.

Lockheed, M. E. (1987, April). *School and classroom effects on student learn-ing gain: the case of Thailand*. Paper presented at the American Educational Research Association, Washington, D.C.

McKnight, C. (1992). *Explanatory notes for the mathematics framework* (Survey of Mathematics and Science Opportunities Research Report Series #41): Michigan State University.

McKnight, C., & Britton, E. D. (1992). *Training Manual, document analysis, in-depth topic trace mapping, regional meetings* (Survey of Mathematics and Science Opportunities Research Report Series #44): Michigan State University.

McKnight, C., Britton, E. D., Valverde, G. A., & Schmidt, W. H. (1992a). *Document Analysis Manual* (Survey of Mathematics and Science Opportunities Research Report Series #42): Michigan State University.

McKnight, C., Britton, E. D., Valverde, G. A., & Schmidt, W. H. (1992b). *In-depth Topic Trace Mapping* (Survey of Mathematics and Science Opportunities Research Report Series #43): Michigan State University.

McKnight, C. C., Crosswhite, F. J., Dossey, J. A., Kifer, E., Swafford, J. O., Travers, K. J., & Cooney, T. J. (1987). *The Underachieving Curriculum: Assessing U.S. School Mathematics from an International Perspective*. Champaign, IL: Stripes Publishing Company.

Ministerio de Educación y Ciencia (1989). Libro Blanco Para la Reforma del Sistema Educativo. Madrid: Martin Alvarex Hnos.

Ministry of Education and Research (1990). Curriculum Guidelines for Compulsory Education in Norway. Aurskog, Norway: Aschehoug

Monchablon, A. (1995). France: System of education. In T. Husen & T. N. Postlethwaite (Eds.), The International Encyclopedia of Education, (pp. 2377-85). Oxford, England: Pergamon.

Moser, U. (1992a). Was wissen 13-jährige in Mathematik und Naturwissenschaften? (What do 13-year-olds know in mathematics and science?). *Schweizerische Lehrerzeitung, 137,* 4-7.

Moser, U. (1992b). Was wissen 13-jährige? Ergenbnisse zu den geprüften Inhalten und kognitiven Fertigkeiten (What do 13-year-olds know? Results divided for content areas and levels of cognitive processing). *Mathematik Bulletin Schweiz, 3,* 1-3.

Moser, U. (1993). The importance of international studies for Switzerland. *The School Field. International Journal of Theory and Research in Education, 1,* 75-88.

*Moser, U. (1994). Die Bedeutung der Leistung in Schule und Gesellschaft (The importance of achievement for school and society). *Beiträge-Infromationen der Pädagogischen Hochschule St. Gallen, 6,* 3-10.

Moser, U. (1995). Que savent les élèves? Etude du Système Scolaire Suisse (What do students know? Evaluation of the Swiss School System). *Math Ecole, 34,* 37-38.

*Nergård, T. (1994). *Hvor er det blitt av naturfagene på barnetrinnet? (What happened to science at the primary stage?)*. Unpublished Masters Thesis, University of Oslo.

*Nezel, I. (1994). *Individualisierung und Selbständigkeit (Individualization and Autonomy)*. Zürich: Pestalozzianum Verlag.

NW EDK. (1995). *ELF: Erweiterte Lernformen. Ein Projekt macht Schule (New approaches of teaching. A project for Schools)* . Bern: Nordwestschweizerische   Konferenz   der   kantonalen Erziehungsdirektoren (NW EDK).

O'Day, J. A., & Smith, M. S. (1993). Systemic educational reform and educational opportunity. In S. H. Furhman (Ed.), *Designing coherent educational policy*, . San Francisco: Jossey-Bass.

*Oakes, J., & Guiton, G. (1995). Matchmaking: The dynamics of high school tracking decisions. *American Educational Research Journal, 32*, 3-33.

*Paulsen, A. C. (Ed.). (1993). *Naturfagenes Pædagogik. Nordic Research Symposium (Science Education. Nordic Research Symposium).* Samfundslitteratur, Denmark: Nordic Research Symposium.

Pelgrum, W. J., & Plomp, T. (Eds.). (1993). *The IEA study of computers in education : implementation of an innovation in 21 education systems.* Oxford, England/New York: Pergamon Press.

Peterson, P. L. (1990). Doing more in the same amount of time: Cathy Swift. *Educational Evaluation and Policy, 12*, 261-280.

Porter, A. C. (1991). Creating a system of school process indicators. *Educational Evaluation and Policy Analysis, 13*, 13-29.

Postlethwaite, T. N., David E. Wiley, with the assistance of Yeoh Oon Chye, William H. Schmidt, & Wolfe., R. G. (Eds.). (1992). *The IEA study of science II : science achievement in twenty-three countries.* (Vol. 9). Oxford, England: Pergamon Press.

Prawat, R. S. (1989). Teaching for understanding: Three key attributes. *Teaching and Teacher Education, 5*, 315-328.

Prawat, R. S., Remillard, J., Putnam, R. T., & Heaton, R. M. (1992). Teaching mathematics for understanding: case studies of four fifth-grade teachers. *The Elementary School Journal, 93*, 145-152.

Putnam, R. (1991). *Instructional practices: conceptual model* (Survey of Mathematics and Science Opportunities Research Report Series #10): Michigan State University.

Putnam, R. T., Heaton, R. M., Prawat, R. S., & Remillard, J. (1992). Teaching mathematics for understanding: discussing case studies of four fifth-grade teachers. *The Elementary School Journal, 93*, 213-228.

Ramsegger, J. (1993). Unterricht zwischen Instruktion und Eigenerfahrung (Teaching between Instruction and Self-experience). *Zeitschrift für Pädagogik, 39*, 825-838.

Ramseier, E., Moser, U., Reusser, K., Labudde, P., & Buff, A. (1994). Schule, Leistung und Persönlichkeit (School, achievement and personality). *Beiträge zur Lehrerbildung, 10*, 67-72.

*Robin, D. (1988, March). *Un marqueur culturel pour l'évaluation comparée des résultats des élèves au niveau international (A cultural marker for the comparative evaluation of results from students at the international level).* Paper presented at the OECD, Poitiers.

*Robin, D. (1989). Teachers' strategies and students' achievement. In L. Burstein (Ed.), *The IEA study of mathematics III: student growth and classroom processes*, (Vol. 3, pp. 225-258). New York: Pergamon.

Robitaille, D. F., Schmidt, W. H., Raizen, S., McKnight, C., Britton, E., & Nicol, C. (1993). *Curriculum Frameworks for Mathematics and Science.* (Vol. No. 1). Vancouver: Pacific Educational Press.

Rotberg, I. C. (1990). I Never Promised You First Place. *Phi Delta Kappan, 72,* 296-303.

Salomon, G. (1992). The changing role of the teacher: from information transmitter to orchestrator of learning. In F. K. Oser, A. Dick, & J.-L. Patry (Eds.), *Effective and Responsible Teaching,* (pp. 35-65). San Francisco: Jossey-Bass.

Schiefele, U., & Csikszentmihalyi, M. (1994). Interest and the quality of experience in classrooms. *European Journal of Psychology of Education, 9,* 251-270.

Schmidt, W. H. (1992). The distribution of instructional time to mathematical content: One aspect of opportunity to learn. In L. Burstein (Ed.), *The IEA Study of Mathematics III: Student Growth and Classroom Processes,* (Vol. 3, pp. 129-145). New York: Pergamon.

Schmidt, W. H., & McKnight, C. C. (1995). Surveying educational opportunity in mathematics and science: An international perspective. *Educational Evaluation and Policy Analysis, 17,* 337-353.

Schmidt, W. H., Porter, A. C., Floden, R. E., Freeman, D. J., & Schwille, J. R. (1987). Four patterns of teacher content decision making. *Journal of Curriculum Studies, 19,* 439-455.

Schmidt, W. H., Putnam, R., & Prawat, R. (1991). *Conceptual Frameworks: OTL, Instructional Practices, Teacher Knowledge and Beliefs* (Survey of Mathematics and Science Opportunities Research Report Series #9): Michigan State University.

Schwille, J., & Burstein, L. (1987). The Necessity of Trade-Offs and Coalition Building in Cross-National Research: A Critique of Theisen, Achola, and Boakari. *Comparative Education Review, 31,* 602-611.

Schwille, J., Porter, A., Belli, G., Floden, R., Freeman, D., Knappen, L., Kuhs, T., & Schmidt, W. (1983). Teachers as policy brokers in the content of elementary school mathematics. In L. S. Shulman & G. Sykes (Eds.), *Handbook of Teaching and Policy,* (pp. 370-391). New York: Longman.

Shulman, L. S. (1986a). Paradigms and research programs in the study of teaching: A contemporary perspective. In M. C. Wittrock (Ed.), *Handbook of Research on Teaching,* (3rd ed., pp. 3-36). New York: Macmillan.

Shulman, L. S. (1986b). Those who understand: Knowledge growth in teaching. *Educational Researcher, 15,* 4-14.

Simons, P. R. J. (1992). Lernen, selbständig zu lernen-ein Rahmenmodell (Learning to learn independently). In H. Mandl & H. F. Friedrich (Eds.), *Lern- und Denkstrategien,* . Göttingen/Toronto/Zürich: Hogrefe.

*Sjøberg, S. (1993). *Naturfag og norsk skole. Elever og lærere sier sin mening (Science in Norwegian schools. Teachers and pupils express their meaning).* Oslo: Universitetetsforlaget.

*Sjøberg, S., & Jorde, D. (1995). Educational reforms in Norway: improving the status of school science? *International Journal of Science Education, 17,* 519-529.

*Sjøberg, S., Jorde, D., Haldorsen, K., & Lea, A. (1995a). *Naturfagutredning rapport 1 (National Science Review, report 1)* . Oslo, Norway: Ministry of Education.

*Sjøberg, S., Jorde, D., Haldorsen, K., & Lea, A. (1995b). *Naturfagutredning rapport 2 (National Science Review, report 2)* . Oslo, Norway: Ministry of Education.

Sosniak, l., & Perlman, C. (1990). Secondary education by the book. *Journal of Curriculum Studies, 22,* 427-442.

Stebler, R., Reusser, K., & Pauli, C. (1994). Interaktive Lehr-Lern-Umgebungen (Interactive Teaching Environments). In K. Reusser & M. Reusser-Weyeneth (Eds.), *Verstehen,* (pp. 229-257). Bern: Huber.

Stigler, J. W., & Perry, M. (1988). Mathematics learning in Japanese, Chinese, and American classrooms. In G. B. Saxe & M. Gearhart (Eds.), *Children's Mathematics,* (Vol. 41, pp. 27-54). San Francisco: Jossey-Bass.

Stodolsky, S. S. (1988). *The subject matters: classroom activity in math and social studies.* Chicago: University of Chicago.

Stodolsky, S. S., & Grossman, P. L. (1995). The impact of subject matter on curricular activity: an analysis of five academic subjects. *American Educational Research Journal, 32,* 227-249.

Survey of Mathematics and Science Opportunities. (1992a). Mathematics Curriculum Framework (Research Report Series #38). East Lansing, MI: Michigan State University.

Survey of Mathematics and Science Opportunities. (1992b). Science Curriculum Framework (Research Report Series #37). East Lansing, MI: Michigan State University.

Survey of Mathematics and Science Opportunities. (1993a). *A Description of the TIMSS' Achievement Test Content Design: Test Blueprints* (Research Report Series #59). East Lansing, MI: Michigan State University.

Survey of Mathematics and Science Opportunities. (1993b). *TIMSS Curriculum Analysis: A content analytic approach* (Research Report Series 57). East Lansing, MI: Michigan State University.

Survey of Mathematics and Science Opportunities. (1993c). *TIMSS: concepts, measurements and analyses* (Research Report Series 56). East Lansing, MI: Michigan State University.

Survey of Mathematics and Science Opportunities. (1995a). *Analysis of Pre-Selected Country Groups* (Curriculum Analysis Technical Report Series #6). East Lansing, MI: Michigan State University.

Survey of Mathematics and Science Opportunities. (1995b). *Document Analysis: Data Collection and Processing* (Curriculum Analysis Technical Report Series #2). East Lansing, MI: Michigan State University.

Survey of Mathematics and Science Opportunities. (1995c). *Document Analysis: Generation of Analysis Files* (Curriculum Analysis Technical Report Series #3). East Lansing, MI: Michigan State University.

Survey of Mathematics and Science Opportunities. (1995d). *Document Analysis: Country Document Profiles* (Curriculum Analysis Technical Report Series #4). East Lansing, MI: Michigan State University.

Survey of Mathematics and Science Opportunities. (1995e). *General Topic Trace Mapping: Data Collection and Processing* (Curriculum Analysis Technical Report Series #5). East Lansing, MI: Michigan State University.

Survey of Mathematics and Science Opportunities. (1995f). *The TIMSS Curriculum Analysis: Overview of an Integrated System of Curriculum Measurement* (Curriculum Analysis Technical Report Series #1). East Lansing, MI: Michigan State University.

Theisen, G. L., Achola, P. P. W., & Boakari, F. M. (1983). The Underachievement of Cross-National Studies of Achievement. *Comparative Education Review, 27*, 46-68.

Thomas, A. J. (1992). Individualised Teaching. *Oxford Review of Education, 18*, 59-74.

Thompson, A. (1992). Teachers' beliefs and conceptions: A synthesis of the research. In D. A. Grouws (Ed.), *Handbook of Research on Mathematics Teaching and Learning*, (pp. 127-146). New York: Macmillan.

Tyack, D., & Tobin, W. (1994). The "grammar" of schooling: Why has it been so hard to change? *American Educational Research Journal, 31*, 453-479.

Valverde, G. A. (1993, March). *Issues in the comparative study of curricula: A review of the literature and directions for further research.* Paper presented at the Comparative and International Education Society, Kingston, Jamaica.

Valverde, G. A. (1995). The United States of America: System of education. In T. Husen & T. N. Postlethwaite (Eds.), The International Encyclopedia of Education, (pp. 1033-41). Oxford, England: Pergamon.

VanLehn, K. (1989). Problem Solving and Cognitive Skill Acquisition. In M. I. Posner (Ed.), *Foundations of Cognitive Science*, (pp. 527-579). Cambridge, MA: MIT Press.

*Walker, D. A., Anderson, C. A., & Wolf, R. M. (Eds.). (1976). *The IEA six subject survey: an empirical study of education in twenty-one countries*. (Vol. 9). New York: Wiley.

Wiley, D. E., Schmidt, W. H., & Wolfe, R. E. (1992). The science curriculum and achievement. In T. N. Postlethwaite & D. E. Wiley (Eds.), *The IEA Study of Science II: Science Achievement in Twenty-Three Countries*, (pp. 115-124). New York: Pergamon.

*Wiley, D. E., & Wolfe, R. E. (1992). Opportunity and achievement: What they tell us about curriculum. In T. N. Postlethwaite & D. E. Wiley (Eds.), *The IEA Study of Science II: Science Achievement in Twenty-Three Countries*, (pp. 115-124). New York: Pergamon.

CONTRIBUTORS

Emilie Barrier is Deputy Head of the Department on Educational Systems at the Centre International d'Etudes Pédagogiques (CIEP) in Paris. Having earned degrees in physics and chemistry from the Sorbonne, she taught physical sciences for over fifteen years at the high school level. As a researcher with the Institut National de Recherche Pédagogique she has participated in many evaluation and comparative education studies. She has also served as the Co-National Research Coordinator for France's participation in several IEA studies including the Second International Mathematics Study, the Reading Literacy study, Computers in Education, and the Foreign Language Study. A frequent consultant to education systems in several countries, she specializes in the audit and evaluation of educational systems and comparative education.

Edward D. Britton, Associate Director of the National Center for Improving Science Education (NCISE), recently contributed to and edited *Examining the Examinations: An International Comparison of Science and Mathematics Examinations for College-Bound Students* published by Kluwer. As part of a multinational project of the Organization for Economic Co-operation and Development (OECD), he is coordinating case studies of eight major US innovations described in *Bold Ventures: US Innovations in Science and mathematics Education* , a three-volume set from Kluwer. He also has written on indicators for science education, dissemination of innovations, curriculum studies, and evaluation. Britton has managed the development of CD-ROM disks and videotapes designed to help teachers enhance their science knowledge and pedagogy.

Leland Cogan, Senior Researcher with the US National Research Center for TIMSS at Michigan State University, coordinated data collection and analysis for SMSO and the TIMSS context questionnaires pilot. Together with William Schmidt, he co-authored the chapter on the TIMSS Context Questionnaire Development for the TIMSS Technical Report volume. His research interests include parent's and teacher's beliefs; students' self-regulation, learning, and motivation; and educational practices. He is currently a member of the team working on the TIMSS international curriculum analysis and the US TIMSS survey.

Doris Jorde is Associate Professor in Science Education at the University of Oslo, Norway. Her research areas include science curriculum studies; primary science; museum education; and international comparative studies in science and mathematics. She has currently been involved in writing the national science curriculum for Norway for grades 1-10. In addition to participation in this study, she conducted the Norwegian science case study for the Organization for Economic and Co-operation and Development (OECD) project in innovations in science and mathematics education.

Ignacio Gonzalo is Professor and Director of the Instituto de Ciencias de la Educación at the Universidad Pontificia Comíllas in Madrid, Spain. Formerly a primary teacher of mathematics and the sciences, he was Head of the Curriculum Innovation Service in the Ministry of Education and Science of Spain which was the project that lead to the implementation of the new syllabi for Infant, Primary, and Secondary Education within the framework of the educational system Reform Act. Evaluation, curriculum design and in-service teacher training strategies are his main fields of interest.

Curtis McKnight is currently Professor of Mathematics at the University of Oklahoma where he has been since 1981, serving as Associate Chairman of the Mathematics Department from 1987 to 1994. He has his PhD from the University of Illinois. He served as National Research Coordinator and Executive Director of the US National Center for the Second International Mathematics Study (SIMS). He serves as the senior mathematics consultant to the US Research Center for TIMSS. He is the author of more than 100 publications and papers. His specialties include cross-national comparative studies, cognitive studies of mathematics learning and performance, and curriculum policy studies such as those involved currently in US calculus reform curricula. He has received over 20 research grants and served as consultant on many other projects.

Urs Moser has been Education Research Scientist at the office of Educational Research, Bern, Switzerland, since 1989. Educated at the University of Fribourg, Switzerland (MA and PhD), he has focused his research on international comparative studies in education, school effectiveness, and the evaluation of different Swiss cantonal educational systems. He also is a research associate with the Institute of Pedagogy at the University of Bern where he is evaluating Swiss cantonal school systems. Moser is a member of the Swiss TIMSS team and is presently engaged in analyzing and publishing the Swiss TIMSS results.

Richard S. Prawat is Chair of the Department of Counseling, Educational Psychology, and Special Education and Professor of Teacher Education. He has been a Senior Research Associate in three federally funded research centers at Michigan State University and is currently involved in the multi-state Educational Policy and Practices Study, examining the effects of current reform initiatives in key sites around the US. He is the author of more than 150 publications and papers in educational psychology. His specialties include constructivist studies of learning and motivation and mathematics and science teaching reform. He served as Associate Director of the Holmes Group from 1985-87.

Senta Raizen is the Director of NCISE. She is the primary author of several NCISE reports on science education in elementary, middle, and high school and books on indicators in science education, preservice education of elementary school teachers, and technology education. Her work also includes educational assessment and program evaluation, education policy, reforming education for work, and linking education research and policy with practice. She is principal investigator for NCISE's evaluations of several federal programs that support science education, the TIMSS study, and the OECD case study project, for which she is working on a synthesis of 23 case studies of innovations in science, mathematics, and technology education. She is a member of the International Steering Committee for TIMSS and serves in an advisory capacity to several national education studies, including the National Assessment of Educational Progress (NAEP), the National Goals Panel, and the National Institute for Science Education.

Prof. Toshio Sawada is a Director of Research Center for Science Education, National Institute for Educational Research located in Tokyo, Japan. He is a member of International Steering Committee of TIMSS. His main interest is Assessment and Evaluation in Mathematics Education, Educational Statistics. He is also a member of International Program Committee of International Congress on Mathematics Education, Spain 1996 and a secretary-general of International Congress on Mathematics Education, Japan 2000.

Katsuhiko Shimizu is Head of Teaching Materials in the Research Department of Teaching and Guidance at the National Institute for Educational Research (NIER) located in Tokyo, Japan. He is currently a member of the TIMSS team at NIER and a member of several curriculum improvement committees under the Ministry of Education, Sports and Culture. He has written in such areas as the use of technology in mathematics education, a psychological perspective on children's mathematics learning, and process aspects of mathematics education.

William H. Schmidt is a Professor of psychometrics and statistics at Michigan State University and the National Research Coordinator and Executive Director of the US National Center which oversees participation of the United States in the IEA sponsored Third International Mathematics and Science Study (TIMSS). He is also a member of the TIMSS International Steering Committee and the TIMSS Technical Advisory Committee. He was a member of the Senior Executive staff and Head of the Office of Policy Studies and Program Assessment for the National Science Foundation in Washington, DC from 1986-1988. He was involved in the IEA's Second International Mathematics Study (SIMS) and the Second International Science Study (SISS) and authored or co-authored chapters in several books reporting on these studies. Much of his recent research and writing has focused on the use of quantitative data to inform public policy particularly in the area of education. His current research focuses on curricular differences and their impact; the distribution and origin of students' learning opportunities; and the effects of curriculum policy. He serves on the AERA Grants Board and is a member of the IEA International Technical Committee.

Gilbert Valverde is Assistant Director of the US National Research Center for the Third International Mathematics and Science Study (TIMSS) at Michigan State University. He coordinated early development of survey instruments and data collections for the SMSO. He is currently a senior member of the team conducting the TIMSS study of mathematics and science curricula in fifty countries. He has written on topics such as the international context of mathematics and science educational policy making, curriculum-sensitive cross-national studies of educational systems, and international development assistance for education.

David E. Wiley is Professor at the School of Education and Social Policy, Northwestern University and Technical Director for the New Standards Project. A statistician and psychometrician by training, much of his recent research and writing has focused on public policy and program evaluations related to educational testing, teaching-learning processes, and legislative initiatives affecting these aspects of education. Having been involved in international comparative education studies since 1971, he recently completed (with T.N. Postlethwaite) a volume reporting findings from the second science study of the International Association for the Evaluation of Educational Achievement (IEA) and serves on the IEA International Technical Committee. He is also working to design and implement new systems based on students' performance on extended response (as opposed to multiple choice) test tasks with the California Learning Assessment System, the state of Kentucky, and the New Standards Project. His current research focuses on the implementation of curricular control policies; the determinants and distribution of learning opportunities; and the integration of frameworks for the assessment of learning, ability, and performance.

Richard G. Wolfe is on the faculty of the measurement and evaluation and computer application programs at the Ontario Institute for Studies in Education at the University of Toronto. He worked on the international design and analysis of IEA's Second International Mathematics Study, on the archiving and analysis of the IEA's Second International Science Study, and on the initial design of the IEA's Third International Study of Mathematics and Science. He has also been involved in initiatives for comparative educational achievement surveys in Latin America. His specialties are achievement survey design and statistical and graphical analysis of educational assessment data.